What people are sayir

The Mysteries of Reality

Gayle Kimball's intelligent and probing interviews become a fascinating excursion, tracking the insights of a remarkable collection of scientists and seekers pushing the limits of what we know. This trilogy will become a prime source for future historians of science. Gayle Kimball has tapped into the experience and wisdom of the world's top researchers working to reveal the subtle but powerful reach of human consciousness.
Roger Nelson, Director of the Global Consciousness Project

For hearts and minds hungry for more knowledge, this trilogy offers a kaleidoscope viewpoint on truth—who we are, why we are here, and how the universe functions. Kimball draws out the healing essence of leaders in the field, weaving together threads of insight through many facets of human experience, like the many faces of a diamond coalescing into the making of healing miracles.
Barbara Stone, Ph.D., author of *Invisible Roots* and *Transforming Fear into Gold*

The Mysteries Trilogy conveys a roadmap to the future! Humanity is continuously redefined by our relationships with each other and the cosmos, and these remarkable testimonies speak of how consciousness provides a key to understanding that the Mind is not limited to the body, but is the pathway to exploring life's infinite possibilities.
Drs. J.J. and Desiree Hurtak, The Academy For Future Science

The Mysteries of Reality is a fascinating book that helps readers see the underlying thought processes of scientists who are expanding our understanding of reality and the nature of consciousness

itself. Interviewer Gayle Kimball asks provocative questions about the circumstances that formed the mindsets and careers of these men and women, and led to their in-depth research on such topics as the true nature of time, psi phenomena, and the survival of consciousness.
Marjorie Woollacott, Ph.D., President, the Academy for the Advancement of Postmaterialist Sciences (AAPS)

Dr. Gayle Kimball has succeeded in bringing together a great variety of viewpoints, in a variety of fields, all lively while deep, through her interviews of many professionals. The Mysteries Trilogy has taken a lot of effort, energy and vision. I am sure they will remain not just a testament to emerging truths but also an amazing way to gather information, belief systems, and everyday life stories of many remarkable people, while keeping high professional standards. She herself is a remarkable individual and the hard work she carried out shows in the writing.
Menas Kafatos, physics professor and author

As we stand at the threshold of an emerging paradigm in which consciousness becomes fundamental in science, historians and philosophers of science will find these words and thoughts essential to understanding how that paradigm shift occurred.
Stephan A. Schwartz

Getting scientists of any ilk to admit to anything mystical or spiritual in their lives is quite a task—as you might suspect since scientists are so logically and usually materialistic-minded in their views. Gayle Kimball has done just that, and it makes for some interesting reading as to what drives the scientific inquiry. I'll give you a hint: It's not logical.
Physicist Fred Alan Wolf, Ph.D., aka Dr. Quantum®

How often are we privileged to enter the minds and hearts of visionary scientists working at the cutting edge of exploration? Dr. Kimball offers this gift to all of us, including the scientists who participated in the dialogues that deserve to be celebrated and widely read.
Gary E. Schwartz, Ph.D., Director of the Laboratory for Advances in Consciousness and Health at the University of Arizona. His recent books include *Super Synchronicity, Is Consciousness Primary?,* and *Extraordinary Claims Require Extraordinary Evidence*

Throughout history, there have been visionaries able to "see" further than others. Many have been spiritual mystics or those with great dreams and the will to pursue them into manifestation. Some have been scientific pioneers including Einstein, Planck, Tesla, Wheeler and Bohm. In this timely, far reaching and deeply insightful trilogy, the visionary clear-sight and wisdom of 65 leading-edge scientists and in their own testimonies are remarkably synergized by Gayle Kimball. They reveal a visionary science seeing beyond a material worldview, based ultimately on the oneness of all creation and the naturalizing of non-local awareness and supernormal phenomena. And founded in an emergent and yet ancient perception that mind and consciousness aren't something we have but what we and the whole world are.
Jude Currivan, Ph.D., cosmologist, healer, futurist, author of *The Cosmic Hologram* and co-founder WholeWorld-View

This informative trilogy offers a treasure trove of personal reflections and insights from many of the leading figures in the modern scientific awakening to deeper understanding of consciousness, the brain-mind connection, and the ultimate healing force of our free will.
Eben Alexander, M.D., neurosurgeon and author of *Living in a Mindful Universe* and *Proof of Heaven*

The Mysteries of Reality

Dialogues with Visionary Scientists

The Mysteries of Reality

Dialogues with Visionary Scientists

Gayle Kimball, Ph.D.

IFF BOOKS

Winchester, UK
Washington, USA

JOHN HUNT PUBLISHING

First published by iff Books, 2021
iff Books is an imprint of John Hunt Publishing Ltd., No. 3 East Street, Alresford,
Hampshire SO24 9EE, UK
office@jhpbooks.com
www.johnhuntpublishing.com
www.iff-books.com

For distributor details and how to order please visit the 'Ordering' section on our website.

ISBN: 978 1 78904 530 7
978 1 78904 531 4 (ebook)
Library of Congress Control Number: 2020930636

A CIP catalogue record for this book is available from the British Library.

Design: Stuart Davies

UK: Printed and bound by CPI Group (UK) Ltd, Croydon, CR0 4YY
Printed in North America by CPI GPS partners

Contents

Books by Gayle Kimball

The other two books in the Mysteries Trilogy:
The Mysteries of Healing: Dialogues with Doctors and Scientists (Waterside Press)
The Mysteries of Knowledge Beyond Our Senses: Dialogues with Courageous Scientists (Equality Press)

Essential Energy Tools: How to Develop Your Clairvoyant and Healing Abilities illustrated with videos and CDs (Equality Press)
50/50 Marriage (Beacon Press)
50/50 Parenting (Lexington Books)
(Ed.) *Women's Culture* (Scarecrow Press)
(Ed.) *Women's Culture Revisited* (Scarecrow Press)
The Teen Trip: The Complete Resource Guide (Equality Press)
(Ed.) *Everything You Need to Know to Succeed After College* (Equality Press)
How to Survive Your Parents' Divorce (Equality Press)
Answers to Kids' Deep Questions in Photos (Equality Press)
Your Mindful Guide to Academic Success: Prevent Burnout (Equality Press)
Ageism in Youth Studies: Generation Maligned (Cambridge Scholars)
How Global Youth Values Will Transform Our Future (Cambridge Scholars)
Brave: Young Women's Global Revolution (Volumes 1 and 2, Equality Press)
Resist! Goals and Tactics for Changemakers (Equality Press)
Quick Healthy Recipes: Literacy Fundraiser (Equality Press)
Young Women Climate Activists (in process)

The most beautiful thing we can experience is the mysterious.
It is the source of all true art and science.
Albert Einstein

We ourselves are a part of the mystery that we are trying to solve. ... We cannot get behind consciousness.
Max Planck

Introduction

Scientists Discover Reality is Consciousness

Many of our foundational scientific beliefs are limited in that they can't account for the evidence that consciousness exists beyond the brain. The dominant scientific materialist paradigm denies the power of spirit, the miracles of mind and feelings over matter, unconscious access to information, and the possibility of other dimensions beyond what our physical senses tell us—including life after death. The dominant worldview is like the *Flatland* novel published in 1884 about a two-dimensional world. When the hero first discovers a three-dimensional place, he is only able to see a flat circle. When he sees more and reports back on his discovery, he's persecuted. It's also similar to the story of *The Emperor's New Clothes* where the crowd applauds the lie that the naked emperor is wearing beautiful new robes, until a small boy has the courage to speak the truth. The dominant worldview often ridicules those with the vision to see other dimensions beyond Flatland, while visionary scientists see the denial of anything but the physical as dogma that ignores extensive research and inhibits our access to subtle information.

Courageous scientists, physicians and psychologists are creating a new non-materialist scientific paradigm as they struggle to define consciousness. Professor Chris Roe* explained, "We don't have the remotest idea of what we mean by consciousness," but "working on the fringes of the anomalies starts to give us answers." He has the courage to put his head "above the parapet and say the evidence is quite strong," so he has moved from trying to prove psi to researching the process of how it works. (*An asterisk indicates that the scientist is featured in a chapter in the trilogy.) Shamini Jain* defines consciousness as the "source and substrate of creation. It is beyond mind, emotion and the physical. It's what gives rise to the physical." John Ryan says it's a type of energy.

The visionary scientists say consciousness doesn't originate in the brain and use synonyms like One Mind, spirit, energy, the Force, matrix, hologram, meaning fields, biofield, cosmic intelligence, information, panpsychism, beyond space and time, substrate, non-physical web, spiritual computer, and fifth dimension. Its basic language is mathematics and patterns like fractals. This implies there is another dimension beyond space and time, which may enable psi phenomena like clairvoyance, precognition, and distant healing.

The scientists liken the new understanding of reality to the huge effect of the Copernican revolution that challenged the belief that our earth was the center of the solar system (although about one-quarter of Americans still don't know this, according to National Science Foundation surveys). These cutting-edge trailblazers persist despite denial from those who don't bother to read their scholarly research studies with highly significant statistical results.

Recognition of the importance of our intent and emotions has major implications for health care, morality, and goal achievement. Bernardo Kastrup* views materialist dogma as making us collectively mad, part of a collective trance that leads to immorality. The condition of our environment is proof of this insanity. Larry Dossey* observes we face a "horrible crisis in ethics and morality." Madonna's 1984 song *Material Girl* spells out the belief system that Mister Right is the man with "cold hard cash."

I interviewed 65 visionary scientists, leaders in the Consciousness Movement, to learn what their research findings reveal about reality, as well as discover what about their personal development encouraged them to be so brave. The three books in the Mysteries Trilogy explore how consciousness shapes reality, enables healing and provides access to knowledge from beyond the physical senses. These visionary scientists conclude that we are more than our physical bodies with more potential than we realize and that science needs to expand to account for consciousness. "The new science embraces all of science, but it's reframing it to have a bigger perspective. This means that we can make sense of data that doesn't fit and we can make new predictions that can be tested, confirmed or disconfirmed. It's an expansion of science that reverses and changes the way you experience what you know," explained Gary Schwartz.* He added, "This perspective turns everything on its head. It's like realizing that the sun no longer revolves around the earth," and encourages reuniting science and spirituality, as discussed in Edward Kelly et al., *Beyond Physicalism: Toward Reconciliation of Science and Spirituality*. These scientists are concerned that the prevailing materialist paradigm limits our abilities and has devastating consequences, including climate change.

Reality is not what we think it is—that's what I learned from the visionary scientists.[1] Common sense erroneously tells us that we live in a solid material world, that atoms are like billiard balls with definite locations and time is a one-directional arrow. This is all false. As physicist Max Planck said in 1931, "I have spent my entire life studying atoms and molecules and I'm here to tell you that they don't exist." He explained in his book *The New Science* that there is no matter as such because atoms vibrate and are held together by a force that indicates a "conscious and intelligent mind. This mind is the matrix of all matter," thus matter is derived from consciousness. The materialist belief is that all information comes from the physical senses via the brain although many people experience ESP, telepathy, a precognitive dream or intuition, awareness of being stared at, or a dramatic near-death experience (NDE) that reveals other dimensions, as it did for neurosurgeon Eben Alexander.* Animals also have these abilities, as evidenced in Rupert Sheldrake's research on dogs that are aware of when their person will come home, even at an unexpected time.[2]

The materialist model of science has, of course, produced a great deal, as evidenced in technologies that can send a person to the moon or create artificial intelligence, but it hasn't succeeded in learning very much about our physical surroundings. Over 95% of the universe is invisible dark matter and dark energy that repels gravity.[3] These mysterious forces have been measured and their effects described but not understood. Various theories try to explain gravity but none are definitive. The observable universe made of energy and matter comprises less than 5% of the universe. An interpretation of quantum physics predicts multi-universes beyond the known universe that remain a mystery to us.[4]

Neither do we know much about the earth under our feet, revealed by Robert Macfarlane in *The Hidden Depths of the Underland* or the 2019 film *Fabulous Fungi*. The mycelium that rises to the surface as mushrooms are surprisingly intelligent in that they solve problems, just as slime mold does. The "secret lives" of trees and how they communicate is revealed by biologist Monica Gagliano in *Thus Spoke the Plant*. The classic *Secret Lives of Plants* by Peter Tompkins and Christopher Bird also reveals mysteries around us.

Similarly, genes with a known coding function make up only about 1.5% of our DNA structure, while the non-coding genes are called "junk" and dismissed as useless.[5] In regards to their function, computational biologist Ewan Birney said, "It's slightly depressing as you realize how ignorant you are [about DNA]. But this is progress. The first step in understanding these things is having a list of things that one has to understand, and that's what we've got here."[6] Biologists are learning about epigenetics, discovering that gene expression isn't fixed but changes in response to our emotions and our environment. Our limited knowledge applies to the whole scientific belief system. William Bengston* concludes, "I can tell you there is nothing more liberating than realizing everything you think is true is wrong."

The Visionary Scientists

To find out about the creation of a scientific paradigm that acknowledges we have access to more dimensions than the Flatlanders, I interviewed Ph.D.s and M.D.s who write books and articles that contribute to our understanding of reality and consciousness. This book includes 19 of the scientists; the others interviewed for the trilogy are represented in the book on the mysteries of healing and one on knowledge beyond the senses. I was inspired by attending the Canadian Energy Psychology conference in Toronto in 2018 where I was struck by the unusual coupling of highly educated scientists talking about spirit and personal experiences of the paranormal. As a feminist academic who does clairvoyant work, I find them very intriguing.

The trilogy explores the scientists' life stories and the influences on their beliefs and personal courage, along with their understandings of our purpose as humans and how the universe really works. I identify with what Larry Dossey* pointed out: "It takes a rabble-rouser to actually develop the courage" to take on the establishment. He added that most peer-reviewed medical journals won't touch papers on topics like the efficacy of prayer in healing. Like these visionary scientists, curiosity and truth-telling motivates me to point out that the Emperor is naked; there are multiple dimensions rooted in consciousness and we have access to information and guidance from beyond the five physical senses.

Using snowball methodology, I asked each of the visionary scientists I interviewed to suggest others. I knew of some whom I researched for my book *Essential Energy Tools: How to Develop Your Clairvoyant and Healing Abilities*. Others were speakers featured at conferences that I attended: Science and Consciousness, the International Society for the Study of Subtle Energies and Energy

Medicine, and Energy Psychology. Our video interviews were conducted on Skype and most posted on my YouTube channel for you to see in entirety. (A few interviewees didn't want their videos made public.) The generic questions are listed on the book webpage. Each scientist and I then edited the written transcript to forge a chapter in the trilogy. My questions and comments are in italics, including references to other chapters that address similar themes. Each chapter starts with questions to ponder to organize your thinking about the deep topics.

I attempted to interview psi skeptics including Chris French, Richard Wiseman, Arthur Reber and James Alcock, as well as the Skeptical Inquirer organization, but only Susan Blackmore* accepted. Reber and Alcock responded to a well-documented article on evidence for psi (including ESP and PK) by Etzel Cardeña by stating they didn't read the studies because they knew they violated the laws of physics, i.e., pigs can't fly so why study flying pigs?[7] They're forgetting quantum physics, as discussed by Steve Taylor.*[8] Psi researchers point out that replication is a problem for many scientists, so they rely on meta-analyses of many studies. Physicist Ed May, who worked on the Star Gate program, described himself in an email as a "total physicalist," and provided links to his psi research papers (see endnote 8). He suggested the Laboratories for Fundamental Research website, which he founded to study "anomalous cognition." Some of the trilogy scientists respond to the skeptics in their book *Skeptical About Skeptics*.

Common themes that surfaced about these pioneers indicate that the visionary scientists are highly intelligent and did well in school and university, as you would expect. The majority were first-born in their families (35 compared to 26 latter-born). Most of the US scientists grew up on the East or West Coast or live there now, with a few exceptions such as some born in the Midwestern states like Ohio. Others grew up in the UK or Canada and one each in Germany, both in Brazil and the Netherlands (Kastrup),

Italy, and Greece. Some are first generation with parents from India, Latvia, Palestine, or Ireland, demonstrating the rich contributions provided by immigrant families. Some had health or family problems in their youth that motivated their search for understanding.

Most psi researchers are older white men, as is true of the trilogy authors. John Kruth* reports that when he started doing research at the Rhine Research Center at age 48, he was considered a youngster. Of 60 scientists included in the trilogy, only two are people of color and only 18 are women. The women have fewer children than the 1.9 average per woman in the US and UK, with 16 children total.

Some visionary scientists had unexpected and transformative mystical experiences, like physicians John Ryan* and Richard Moss,* physicist Jude Currivan,* psychologist Steve Taylor,* and linguist J.J. Hurtak.* For a few, an NDE was transformative, such as for Eben Alexander,* Joyce Hawkes, and Marilyn Mandala Schlitz.* Family influences included visionary mothers such as James Carpenter,* David Lorimer,* Christine Simmonds-Moore,* Judith Swack* and John Kruth,* or an influential sibling like David Muehsam,* Marjorie Woollacott,* Mary Rose Barrington,* and Larry Burk.* Psychedelic drugs influenced scientists like Susan Blackmore* and David Luke.* Having his "bad back" healed by a man he met while lifeguarding at a swimming pool led William Bengston* to research healing and an injury led Richard Hammerschlag* to study acupuncture. Others read psi research as teens or were influenced by philosophers such as Teilhard de Chardin or physicist David Bohm. Overall, curiosity about reality data was the main motivation to follow the evidence, however much in conflict with the materialist belief system.

The scientists are often spiritual rather than religious and many went through an adolescent rebellion against religious dogma. More of them have a Jewish background than would be predicted by the small percentage of Jews in the world population. Jeffrey Mishlove* interviewed over a thousand visionary scientists for his video show "New Thinking Allowed."[9] He reported, "I've discovered that quite a number of prominent people in parapsychology have a Jewish background. Jews are a tiny minority but Jewish people are prominent in every cutting edge activity in which I've ever been involved. I would say 95% of the people who explore these areas scientifically do so because, like me, they've had powerful personal experiences. The other 5 to 10% do it out of intellectual curiosity." Les Lancaster explained, "The combination of the value placed on learning, the leanings towards mysticism, the eschatological idea of promoting a golden age to come, and the pressure of being in exile was hugely formative for Jews entering the modern era. This is a potent mix that breeds pioneers!"

The visionary scientists are intuitive types rather than sensing personality types on the Keirsey and Bates scale (available online to compare your scores with the scientists).[10] Only three men scored sensing rather than intuitive. They're more extroverted than the typical research scientist: Our group profile is Extrovert, Intuitive, Feeling, and very close but slightly more Judging, called "Idealist Teachers or Champions." Some are interested in the Enneagram as a tool for self-understanding, like Charles Tart* and Judith Swack.* (More about their typologies is on the book website.)

Some readers wondered why I included astrological types in a book about science: Their most common signs are Sagittarius (10), Aquarius and Libra (both 8). One reason is astrology is a shorthand to personality type for those who find it useful and, second, I was curious how the interviewees would respond to a controversial topic. I've found my natal chart accurate: For example, I have fiery Mars strengthened by its position in the constellation Aries the ram in my 10th house of occupation, indicating I focus on work. The visionary scientists enjoy their research and get grounded by being in nature. An unusual number of them are musicians or singers and some write novels or poetry. "Curious" is the most common word they use to describe their drive to understand reality on a deep level and they enjoy being on the cutting edge.

The New Non-Materialist Paradigm

Although researchers like Dean Radin* report on thousands of double-blind studies with results in some cases trillions of times beyond what you would expect by chance, the findings are often dismissed as pseudoscience or "woo-woo." Radin suggests, "What is needed for a new paradigm is a more comprehensive model of reality where consciousness becomes just as fundamental, if not more so, than materialism." Actually, idealism, defined as the belief that reality is mentally constructed, is very old. Ancient Greek philosophers, such as Anaxagoras in 480 BCE, emphasized the primacy of mind and consciousness, although his contemporary Plato's idea of ideal forms is better known. Ancient Hinduism, Buddhism and Jainism also recognized that the material world is *maya,* illusion. Through meditation, sages develop psi abilities called *siddhis,* such as clairvoyance or bilocation. More modern siddhis are described in *Autobiography of a Yogi* by Paramahansa Yogananda. The Bible is full of references to healing, prophecy, prayer, and what Paul calls "gifts of the Holy Spirit" in 1 Corinthians 12.

Many of our scientists were influenced by Eastern religious thought and are meditators, while some participate successfully in their own experiments with ESP, etc. They realize the impact of experimenter expectations and value qualitative research such as case studies—including their own first-person psi experiences—what Gary Schwartz* calls "self-science." They encourage openly sharing data online.

Psychic remote viewers, led by Russell Targ* in the Star Gate government programs during the Cold War, sketched secret

Soviet missile silos, a huge new secret Soviet submarine, and the location of a downed Russian military plane before it was discovered by the Soviets in North Africa, and drew targets before they were even selected. The remote viewers were only given numbers with no hint about the target, paid attention to their perceptions and drew pictures of the target while in their offices. Stephan Schwartz* worked with remote viewers to discover archeological finds such as Cleopatra's Palace and the Lighthouse of Pharos––one of the seven wonders of the ancient world––as well as sunken ships and a lost Mayan temple. Working in the military as a remote viewer, Paul H. Smith* found the location of drug contraband in a huge container ship while working in his office. Charles Tart* suggests that remote viewers also could be used in therapy to explore the roots of psychological problems. Medical intuitive Caroline Myss' well-known work with Dr. Norm Shealy in his medical practice can be considered a form of remote viewing of a stranger's body.

It follows from materialism that death ends our awareness, but rigorous triple-blind experiments with mediums who accurately communicate with disembodied spirits are conducted by Ph.D.s Julie Beischel* (Windbridge Institute), Gary Schwartz* (University of Arizona), Chris Roe* (University of Northampton) and others. Over 2,500 well-documented cases of children who remember their past lives were collected by Ian Stevenson and Ed Kelly at the University of Virginia.[11] Over one-third of the children had birthmarks and/or phobias representative of death traumas such as wounds that caused their past deaths, or fear of water caused by a drowning death. Millions of people with NDEs often report undergoing a life review with a loving being where they feel the impact of their lifetime actions.[12] Some of the visionary scientists had communication with a dead relative, such as Russell Targ* whose daughter Elisabeth gave a message of love in Russian so he would know it was her and Fred Alan Wolf's* son communicated with him after being killed in an accident. Denying the long-term consequences of our actions (karma) leads to overemphasis on the worldview of "eat, drink and be merry for tomorrow we die."

The old view is that time only moves forward, although Albert Einstein explained that space and time are each relative to an observer. It seems like fantasy to think that precognition and retrocausality could exist or that intention in the present may change the past. Yet, Dean Radin's experiments, connecting subjects to physiological measuring devices, while they are shown slides, found that their bodies reacted even before an alarming or arousing slide was selected by a computer. Even more astounding, the past output of Random Number Generator (RNGs) machines could be changed by intention—but only if the output hadn't been read, as Roger Nelson* found in his PEAR laboratory at Princeton. (RNGs are computer devices that generate random 0s and 1s.) Remote viewers accurately draw a target before it's selected, as Russell Targ* describes.

The academic bias against psi research means that it's neglected and underfunded to our detriment, in what Jim Carpenter* calls a huge blind spot. Psi is a Greek letter, the first letter of the word "psyche," meaning mind. Psi is used to refer to anomalous or extraordinary experiences, parapsychology and the paranormal. However, Carpenter argues in *First Sight* that "para" should be dropped because we so often use these often unconscious intuitive abilities to stay safe. Since the materialist paradigm is so limited and harmful, clearly more research is needed to enhance our physical and emotional health.

The old paradigm believes that scientists can be objective in conducting research, but psi researchers recognize the inevitable effect of the intention and beliefs of the experimenters. Garret Moddel* tried to get around this effect by designing an experiment with two RNGs and two computers with no human involvement. The first round achieved the effect he and his students were looking for, as explained in his chapter. When they ran the experiment again later, there was no effect, leading him to realize that what had changed was his attitude and intent, which the machines reflected. Psi research indicates that we have free will, as opposed to the determinism of the materialist model that says all is determined by chemical interactions, etc. Some quantum physicists also believe the determinism of physical laws denies free will. (*See Ruth Kastner's alternative view.**)

The list of psi resources on the book's webpage shows few pertinent college courses to encourage young scholars. A few donors stand out like Henry and Susan Samueli's 2017 gift of $200 million to the University of California at Irvine to research integrative health care such as acupuncture, naturopathy, homeopathy and herbs. The gift evoked criticism in a *Los Angeles Times* article, citing the fear it would "threaten to tar UC Irvine's medical school as a haven for quacks."[13] The article quotes Professor David Gorski as saying the only reason for integrative medicine is "to integrate quackery into medicine." (The Samueli institute also funded Shamini Jain's* research in energy healing published in scientific journals.)

To counter this academic bias, visionary scientists established organizations like the Scientific and Medical Network (SMN) in the UK, and in the US, the Academy for the Advancement of Post-Materialist Sciences, the Society for Scientific Exploration, and the Consciousness and Healing Initiative. SMN member David Lorimer* explained, "The original Network was set up because it still is very unfashionable and perhaps even dangerous to hold a non-material worldview." These organizations also publish and organize conferences. The SMN's Galileo Commission involved over 90 scientific advisors who recently published a report titled "Beyond A Materialistic WorldView—Towards an Expanded Science." A distinguished and well-known professor who turned down my request to be in the trilogy explained, "I agree that rigorous scientific psi research is dismissed, but I do not want to become embroiled in public controversy about that." In contrast this book's scientists have the courage to be embroiled and we'll learn why.

Useful Applications of the Consciousness Paradigm

If our common sense and simplistic notions about material things, locality and distance, time, sources of information, death, and the power of belief are wrong—at least on the quantum level, what are the implications? We need training in how to accurately access non-sensory information and use thoughts and emotions to manifest goals. It's helpful to learn to pay attention to seemingly unrelated synchronistic events, especially what Gary Schwartz* calls "super synchronicity" when we experience six or more related events. They suggest that some form of helpful guidance exists and that we can learn how to access it more deliberately.

Visionary psychoanalyst Carl Jung developed the concept of synchronicity after one of his patients described a dream about a scarab beetle as a similar beetle appeared on the window. Dawson Church gives many more examples in his chapter on synchronicity in his book *Mind to Matter: The Astonishing Science of How Your Brain Creates Material Reality.* He quotes Jung's definition: "Synchronicity is a meaningful coincidence of two or more events, where something other than the probability of chance is involved." As we grow on a spiritual path, it happens more often, suggesting a subtle intelligence at play.

The old paradigm values logic and analytical thinking but is uncomfortable with emotions and the content of the unconscious, while the multidimensional approach includes what intuition and dreams reveal from the deeper mind. Carpenter points out that materialistic psychology ignores our spirituality. Larry Burk* writes about people who accurately diagnosed their own diseases, such as cancer, by paying attention to their dreams. Getting their doctors to pay attention to these dreams was often problematic, or actually led to death that perhaps could have been avoided if the dream diagnosis was taken seriously. Stephan Schwartz* got involved in studying Edgar Cayce's channeled information when a stranger arrived at his house to tell him she dreamed that he should be involved in the Cayce Foundation's ARE (Association for Research and Enlightenment). He went on to study there for five years. Henry Reed* leads Dream Helper Circles where members commit to dream one night in a therapeutic way for a target member of the circle and discuss their dreams in the morning. Many of the scientists had precognitive dreams themselves.

The materialist paradigm believes that consciousness only exists when manifested by a physical body and a brain since the physical is all that exists. Yet, Charles Tart* observes we are more than the personal and can have extraordinary transpersonal experiences, as acknowledged by transpersonal psychologists. The new (but actually ancient) view acknowledges that other dimensions may exist, not subject to limitations of locality and time. Influential psychologist William James warned against "medical materialism" as being too "simple-minded" in explaining mystical experiences of the Christian saints are purely biological.

Neurosurgeon Eben Alexander* was astounded to experience multiple-dimensions when meningoencephalitis caused severe damage to his neocortex (the outer surface of the brain, most related to our human awareness). He experienced similar visions as Robert Monroe, founder of the Monroe Institute, whom he knew nothing about at the time. Because of medical advances, an increasing number of people survive to tell about their NDEs and many report mystical experiences that change their understanding of the long-term consequences of their actions.

In the old paradigm, since only the material world is viewed as real, useful nonphysical communication is often dismissed. This includes ESP, studied by psychiatrist Diane Hennacy Powell,* and telepathy. Many Ganzfeld studies conducted during relaxation or sleep, and in James Carpenter's* men's discussion group, succeeded in influencing what the subjects were thinking about. Hundreds of the Ganzfeld studies are viewed as the "flagship of experimental parapsychology."[14] In the experiments a computer selects a video clip or photo that is later shown to the subject along with three other images and she or he is asked to identify the target image. The accuracy in hundreds of studies is above the 25% rate predicted by chance. As Dean Radin states, "Materialism entails a certain set of assumptions that are perfectly fine for understanding the physical world. But those assumptions (as we understand them today) cannot easily account for all aspects of reality, especially consciousness and psychic phenomena."

The Flatland worldview is that illnesses are cured by drugs and/or surgery, certainly not by prayer from a distance or by healers, although focused intention changes machines, cells, bacteria, seed growth, plasma, photons, etc. Practical applications of the expanded understanding of our abilities include health care: For example, biochemist Joyce Hawkes does effective healing work from a distance, as discussed in our video interview.[15] The health care of the future will utilize treatment modalities from the East and the West, predicts Richard Hammerschlag, co-founder of CHI. Already many medical schools include integrative health, wellness, and spirituality programs, as discussed in the International Congress on Integrative Medicine and Health.

William Bengston's* skeptical students routinely cure mice injected with mammary cancer that kills control mice in about 27 days, while the healed mice live out their normal two-year life span. Interestingly, biology students who are embarrassed to be sitting in a lab with their hands around a cage full of mice don't have the same curative outcomes, indicating that the attitude of the healer makes a difference. In what seems like a resonance effect, some of the mice injected with breast cancer and placed in used healing cages didn't die of breast cancer, even when the healing practice wasn't consciously directed at them.

Bengston is researching duplicating the results by recording healing frequencies while a group of people he trained heal the mice. His goal is to transmit the inaudible frequencies through speakers for healing others. When cancer cells were placed near the speakers in Bengston's lab, 68 significant genomic changes occurred in 167 cancer genes. He reports, "Certainly the cells were able to recognize that something was going on here and they responded."

We've known for a long time about the power of placebo to heal, about spontaneous remission of terminal illnesses, and how a person with Dissociative Identity Disorder with multiple personalities ("alters") can have diabetes with one alter and not the other. One alter or personality is allergic and the other isn't; one alter needs glasses and the other doesn't.[16] They can react differently to medication, have different blood pressure readings, heart rate, and EEG readings, which indicates the influence of mind over matter. Seemingly miraculous occurrences, such as the growing percentages of pharmaceutical trials where the placebo is almost as effective as the drug, are dismissed by researchers as irritating. However, logic indicates they should try to figure out how to stimulate the use of suggestion or belief to heal. Placebo has an impact even if the subject knows it's a sugar pill, especially if the pill is colorful and large, as researched by Harvard Professor Ted Kaptchuk.

Hypnosis can also produce biological effects. For example, telling a subject a pencil touch is a lit cigarette can raise a blister on the skin.[17] Prayer can assist in healing; see Larry Dossey's* chapter for an explanation of a flawed Harvard University study that discounted its effectiveness. Psychologist Chris Roe* points out that the evidence "suggests our current psychological model of what it is to be a human being is incomplete."

Psychiatrists Robert Alcorn* and Mitchell Gibson,* along with therapist and "Soul Detective" Barbara Stone,* were surprised to discover the negative impact of other-dimensional entities on the mental health of their patients. These three therapists proceeded to develop techniques to remove the invading energies, which improved the lives of their patients. They may call on angelic beings and spirit guides for guidance.

Some healers found that the energy field is the template for the physical body, and work with the meridians, chakras, and the auric field. John Ryan, M.D.,* for example, uses chakras in his healing work, as I do in phone sessions, as I view them as subconscious memory banks. I also use energy psychology tapping on meridian points, such as Emotional Freedom Technique. Yet very little research has been invested in discovering how these phenomena work and how to apply them to healing. The National Institutes of Health (NIH) budget for the National Center for Complementary Health (established in 1992) should be much larger. Only $126,081,000 was allocated for fiscal year 2020, compared to $43 billion for the NIH.

These scientists report that factors that may influence psi abilities (like ESP) include: previous experience and belief in psi phenomenon, experienced meditators, creative, fantasy proneness, artists, musicians, relaxed rather than anxious, extroverted, open, able to focus (measured by the Absorption Scale), boundary thinness, positive schizotypy (sometimes called "magical thinking"), bioeccentricity, and being left-handed or ambidextrous. Diane Hennacy Powell adds: genetics, history of severe trauma (especially in childhood), history of an NDE, ADD, Bipolar Disorder, autism, and being a mother (in part because of the brain remodeling that occurs under the influence of hormones during pregnancy).

Business and finance are often interested in future forecasts and could utilize psi research and remote viewing to identify future trends, as in Stephan Schwartz's* work, or as explained in Julia Mossbridge's *The Premonition Code.* She suggests that "precogs" could help NGOs predict famines, help patients, guide new college students, etc. Some remote viewers applied their skills to earning money on the Commodities Market. Patrizio Tressoldi* works with devices that respond to intention, hoping that, "our studies of practical application of mental interaction from a distance are replicated by investigators to convince people that our human potentialities are much more than we experience in normal life." Menas Kafatos* concludes, "The message of all the different spiritual traditions is we are more than we think we are." The materialist dogma limits our abilities to only what's believed to be common sense and dulls awareness of long-term responsibility for our actions.

Quantum Physics is the Foundation of the New Paradigm

Quantum mechanics is often relied on to explain psi phenomena, which makes most physicists uncomfortable. There are a variety of quantum theories such as the Russian scientists' torsion field theory[18] or Ruth Kastner's* development of the Transactional Interpretation. A fascinating hidden universe beyond space-time is revealed by the invisible world of quantum physics. It doesn't follow the classical laws of physics, leading physicist Anton Zeilinger to observe, "The world is even weirder than what quantum physics tells us." When two photons are entangled or bonded, even though they appear to be separate, if one is sent to another solar system and its spin is measured, the other instantaneously reflects it. This non-local connection bothered Albert Einstein who dismissed it as "spooky action at a distance." He also said, "Concerning matter we have been all wrong. What we have called matter is energy whose vibration has been so lowered as to be discernible to the senses. There is no matter."

However, scientists have observed non-local entanglement not just with tiny photons or atoms, but with macroscopic objects like diamond crystals.[19] In 2018 a team from MIT entangled photons of light in a lab with starlight from 600 light years away and light from two distant quasars, the further of which is 12.2 billion light years away. This could explain how psi is possible, perhaps one of the reasons quantum physicist Henry Stapp suggested that non-locality may be the most profound discovery in all of science.

Scientists have no definitive classical physics explanation of this non-local distant connection of entangled pairs. How can seemingly separate and distant particles be so connected since Einstein taught us nothing can move faster than the speed of light? Imants Barušs* suggests the existence of intelligent fields beyond space-time. Some relate this dimension to consciousness or the ground of being. It's very appealing to point to quantum non-locality as an explanation for distant healing prayer, ESP or precognition. However, physicists don't know if there is an information field or what consciousness is or how it could arise from matter. This is referred to as "the hard problem of consciousness" that can't be explained by the materialist paradigm. Some of the visionary scientists think of consciousness as benign and helpful, while others don't attribute feeling to it. They agree it's the ground of being, the source of the material world, not the other way around and that it's difficult to define.

Physics Professor David Kagan warned in a personal communication:

> *If you hang your hat on today's unresolved mysteries of science, you will likely look foolish because eventually these mysteries will be solved and new ones created in the process. Physics is a false god for those pleading their case that their work is valid. Physics changes all the time. It is not The Absolute Truth. It is but one "way of knowing," albeit, it is a powerful way of understanding the structure of our universe from a physical perspective. Nonetheless, it can't be the absolute truth because it is an infinite onion, one layer peeled back at a time revealing yet another layer—ad infinitum, an image suggested by Richard Feynman.*

Some physics theories suggest there are more than three dimensions of space and more than one time dimension. Some scientists believe that the invisible quantum world exists in potentiality and a wave collapses into a particle only when it's observed or measured. Another theory is that multiple states actually exist at the same time in distinct parallel universes, as in the popular Many-Worlds theory of Hugh Everett. Over time, the Copenhagen interpretation of Niels Bohr predominated, stating that a quantum particle potentially (but not actually) exists in all possible states at once. Trying to understand what potentiality is before it's measurable, physicist Fred Alan Wolf* explained to me, "Before collapse, only mind stuff exists in the guise of possibilities. Possibilities are potentially able to be something. Atoms are ideas we use to map out what we observe with sensitive instruments." Quantum physics clearly is weird and spooky, as Einstein said. What we know for sure is that we know very little about our world and therefore should be humble and open to new ideas. As Niels Bohr said to Wolfgang Pauli about his theory of elementary particles, "We are all agreed that your theory is crazy. The question that divides us is whether it is crazy enough to have a chance of being correct."[20]

Theoretical physicist Sean Carroll explains that versions of quantum mechanics agree that the universe is composed of wave functions that when measured collapse into either particles (a "cloud of probability" like an electron or the photons in light) or stay as waves.[21] "The world is wavy," he says. The version that makes more sense to Carroll is Many-Worlds where atoms are in superposition of every possible position until observed, hence there are "many copies of what we think of as the universe." He reports that scientists agree they don't really understand the quantum realm; hence the well-known saying, "Shut up and calculate."

Many people assume that the effect of quantum physics was to open scientists to the mysterious as Einstein suggested in the book's opening quotation. To the contrary, Wikipedia, YouTube, and journal articles still attack or censor our vanguard scientists as pseudo scientists. Wikipedia's first sentence about brilliant researcher Dean Radin* reads: "He has been Senior Scientist at the Institute of Noetic Sciences (IONS), which is on Stephen Barrett's Quackwatch list of questionable organizations." Another example is Larry Burk,* M.D., who reports, "I gave a TEDx talk, which got censored with a disclaimer by TEDx," regarding his research about women whose dreams revealed their breast cancers. Russell Targ's* TED talk about psychic abilities was banned but you can see it on YouTube. It's still rare to get psi articles published in scientific journals or obtain funding for psi research. Although Brian Josephson* is a Nobel Prize-winner in physics, he reports that Cambridge University graduate students are steered away from him by faculty, slowing his research despite what Nikola Tesla predicted: "The day science begins to study nonphysical phenomena, it will make more progress in one decade than in all the previous centuries of its existence."

My Motivation

My interest in the topic is both a personal and academic interest in spirituality, earning a Ph.D. in Religious Studies from the University of California at Santa Barbara. As a teenager, I had my pivotal "ah-ha" experience. The chapter on Hinduism in Huston Smith's book *Religions of Man* (back in the old days when people didn't use inclusive language) explained cause and effect, karma and reincarnation. This gave me the vocabulary to understand the purpose of life as attracting experiences to grow and blossom. When I ask my workshop participants to pick one word to describe them, I often start with "curious." (My typologies are E/INFJ, Enneagram Type 1, Gemini, and first-born.)

I could best understand the notion of an unseen intelligence shaping the important events in my life as similar to physics. When an atom is missing an electron, it attracts the electron it needs to be complete. Very few major events in my life were orchestrated by my conscious mind, except my choice of undergraduate university at UC Berkeley and my first teaching job. I was guided to other milestones since I didn't know enough to ask for them, as when, at a conference, I met professors from California State University, Chico who were looking for their token woman instructor.

I never dreamed I'd switch from teaching high school history, to university Religious Studies, to Women's Studies, and now to teaching clairvoyance and healing and doing individual sessions in countries ranging from Japan to Canada, as well as writing books about global youth. I learned clairvoyant techniques when I was on a sabbatical from teaching and decided I wanted to focus more on spirituality. I'd experienced snatches of psychic information, seeing snapshots of men I would meet in the future and repeating to a boyfriend his thoughts or conversations, but only in anxious moments. I took classes for a year from the Chico Psychic Institute's clairvoyant program. It's useful to be able to turn on my inner vision at will rather than waiting for some emotionally charged situation to turn it on and to be able to assist others to get to their core issues. Researchers debate if psi abilities can be taught: In my experience the answer is yes, it can be taught like most skills, as explained in my *Essential Energy Tools*.

In this book the scientists report on their research findings, as well as how they stay centered and inspired. Meditation or prayer, exercise, being in nature and looking at challenges as opportunities for growth keep them in the flow. I wanted to know how they cope with difficult challenges and also how they enjoy life. Most would agree with what Dean Radin said, "I find it exhilarating to explore the edge of the known," or as Christine Simmonds-Moore reported, "I've always been fascinated by mystery."

Concerned about climate change, growing economic inequality, and the increasing number of autocrats, I asked if they are optimistic or pessimistic. In thinking about the future, some scientists are pessimistic, especially about the harms caused by climate change. Civilization as we know it may not survive, warns Stephan Schwartz. We're the only self-destructive species, says William Bengston. We're in some ways devolving and creating disasters, agrees Mitchell Gibson. Capitalism is messing us up, points out Brian Josephson, and James Carpenter agrees the wealthy elite has too much power. The optimists believe that we're seeing a global awakening of spiritual awareness, especially among young people, and that part of the transition is exposing the broken systems and injustices like racism and sexism. We're seeing an uprising of sanity, believes Larry Dossey. Some pointed to Steven Pinker, author of *The Better Angels of Our Nature,* who argues that violence and wars have decreased. Others believe we're part of a divine plan and have non-human helpers. Advances in technology will be helpful, although side effects such as job loss or too much time spent in front of screens are problematic. The bottom line is will we turn around climate warming? All agree the earth will carry on although human civilization as we know it may not.

I found the interviews fascinating, informative, sometimes funny and always eye-opening, as I hope you will. These scientists conclude that the materialist paradigm is destructive and limits our abilities. It must be replaced by a more accurate, expansive and useful understanding of reality and our abilities.

Resources

Bibliographies about psi research:

Eben Alexander (http://ebenalexander.com/resources/reading-list)

David Luke (https://visionaryscientists.home.blog/2019/03/13/

psi-resources-online-articles-books-compiled-by-david-luke-ph-d-for-his-course-psychology-of-exceptional-human-experience)
Dean Radin (http://deanradin.com/evidence/evidence.htm)
Charles Tart (http://blog.paradigm-sys.com/links-and-resources)

Trilogy webpage: https://visionaryscientists.home.blog

Characteristics of the Visionary Scientists: https://visionaryscientists.home.blog/2019/12/11/characteristics-of-the-visionary-scientists-featured-in-the-mysteries-trilogy/

This introduction:
https://visionaryscientists.home.blog/2019/11/11/what-is-reality-the-new-non-materialist-paradigm/

Interview Questions:
https://visionaryscientists.home.blog/2019/11/11/interview-questions-for-visionary-scientists/

Psi Resources: Journals, Groups, Universities, Conferences, Websites:
https://visionaryscientists.home.blog/2019/11/11/psi-resources-journals-groups-conferences-websites/ (An historic list of paranormal researchers is available online.[22])

Psi Research:
https://visionaryscientists.home.blog/2019/12/11/psi-phenomena-research-by-visionary-scientists-in-the-mysteries-trilogy/

Video interviews: https://www.youtube.com/channel/UCYQz9QMYs2b1R1uAKnMzWQQ

Abbreviations

AI: artificial intelligence
ESP: extrasensory perception
IONS: Institute of Noetic Sciences
MIT: Massachusetts Institute of Technology
NDE: near-death experience
OBE: out of body experience
PK: psychokinesis (the ability to influence a physical system with intention)
PTSD: post-traumatic stress disorder
QM: quantum mechanics
QP: quantum physics
RNG: random number generator computer, also called random event generator
RV: remote viewing
SRI: Stanford Research Institute

Definitions

Entanglement: A pair of particles are bonded so that they influence each other instantaneously from a distance. Harald Walach* expanded this concept to a Generalized Entanglement model.[23]

Non-locality: Action from a distance implying a universal connectivity.

Panpsychism: Everything material has individual consciousness, in opposition to the dualist belief that material and mind are separate, or the materialist belief in only the physical.

Parapsychology: The British referred to this as psychical research, the scientific study of the paranormal such as precognitive dreams, telepathy, mind influencing matter, etc.

Photon: A particle representing a quantum of light or other electromagnetic radiation.

Presentiment/precognition: Knowledge of a future event.

Psi: A Greek letter used to refer to anomalous experiences, parapsychology and the paranormal. Dean Radin's list of psi categories is: Healing at a Distance, Mind-Matter Interaction (PK or psychokinesis), Physiological Correlations at a Distance, Precognition & Presentiment, Survival of Consciousness after Death, and Telepathy and ESP.[24]

Quantum mechanics/physics: The study of subatomic particles like photons and electrons. They don't follow the same rules of physics as the visible world.

Synchronicity: Associated with Carl Jung, who defined it as, "meaningful coincidence of two or more events, where something other than the probability of chance is involved."

Endnotes

1 Thanks to physicists Fred Alan Wolf, David Kagan, Jude Currivan, and Brian Josephson for critiquing this section, as well as Bernardo Kastrup and Marjorie Woollacott, but errors are my own. Thank you to transcribers Manisha Hariharan, Joshua Herrera and Nicole Hobbs and of course to the courageous and kind visionary scientists. Ten science questions and results: https://www.businessinsider.com/science-questions-quiz-public-knowledge-education-2018-5

2 Our video interview: https://www.youtube.com/watch?v=6rsmIcnmISc&t=117s https://www.nationalgeographic.com/science/space/dark-matter/

3 Katia Moskvitch. "Troubled Times for Alternatives to Einstein's Theory of Gravity." *Quanta Magazine*, April 30,

2018. https://www.quantamagazine.org/troubled-times-for-alternatives-to-einsteins-theory-of-gravity-20180430/

4 Ethan Siegel. "The Multiverse is Inevitable, And We're Living It." *Medium*, October 19, 2017. https://medium.com/starts-with-a-bang/the-multiverse-is-inevitable-and-were-living-in-it-311fd1825c6

5 Nessa Carey. "Junk DNA: A Journey Through the Dark Matter of the Genome," 2017. https://www.discovermagazine.com/health/our-cells-are-filled-with-junk-dna-heres-why-weneed-it "Help Me Understand Genetics." NIH. https://ghr.nlm.nih.gov/primer/basics/

6 Stephen Hall. "Hidden Treasures in Junk DNA." *Scientific American*, October 1, 2012. https://www.scientificamerican.com/article/hidden-treasures-in-junk-dna/

7 Reber, A.S. & Alcock, J.E. "Searching for the Impossible: Parapsychology's Elusive Quest." *American Psychologist*, 2019. http://dx.doi.org/10.1037/amp0000486 https://skepticalinquirer.org/2019/07/why-parapsychological-claims-cannot-be-true/

8 Steve Taylor. "Scientism: When Science Becomes a Religion." *Psychology Today*, November 13, 2019. https://www.psychologytoday.com/us/blog/out-the-darkness/201911/scientism

9 https://www.newthinkingallowed.com/Listings.htm

10 https://www.strategicaction.com.au/keirsey-temperamentsorter-questionnaire/Journal of Contemplative Inquiry. https://stats.stackexchange.com/questions/129684/statisticalsignificance-of-birth-month-of-professional-boxers https://personalitymax.com/personality-types/population-gender/

11 https://med.virginia.edu/perceptual-studies/publications/books/ "Dr. Ian Stevenson's Reincarnation Research." https://www.near-death.com/reincarnation/research/ianstevenson.html

12 "Key Facts About Near-Death Experiences." International

Association for Near Death Studies. https://iands.org/ndes/about-ndes/key-nde-facts21.html?start=1
13 Michael Hiltzik. "A $200-Million Donation Threatens to Tar UC Irvine's Medical School as a Haven for Quacks." *Los Angeles Times*, September 22, 2017. https://www.latimes.com/business/hiltzik/la-fi-hiltzik-uci-samueli-20170922-story.html
14 "Ganzfeld." Psi Encyclopedia. https://psi-encyclopedia.spr.ac.uk/articles/ganzfeld
15 https://www.youtube.com/watch?v=aHlXRERp2pc&t=7s
16 http://traumadissociation.com/alters https://www.quora.com/Can-someone-with-multiple-personality-disorderdevelop-different-physiological-characteristics-allergies-for-example-for-different-personalities
17 Gordon Paul. "The Production of Blisters by Hypnotic Suggestion." http://citeseerx.ist.psu.edu/viewdoc/download?doi=10.1.1.484.79&rep=rep1&type=pdf
18 Claude Swanson. "The Torsion Field and the Aura." ISSSEEM. http://journals.sfu.ca/seemj/index.php/seemj/article/view/425/386
19 Edward Frenkel. "The Holy Grail of Quantum Physics on Your Kitchen Table." *Scientific American*, September 27, 2013. https://www.scientificamerican.com/article/the-holy-grail-of-quantum-physics-on-your-kitchen-table-excerpt/
20 Sean Carroll. *Something Deeply Hidden*. Dutton, 2019.
21 Ibid.
22 https://psychicscience.org/researchers.aspx
23 Karl Tate. "How Quantum Entanglement Works." *Live Science*, April 8, 2013. https://www.livescience.com/28550-how-quantum-entanglement-works-infographic.html
24 http://deanradin.com/evidence/evidence.htm

Section 1

New Scientific Paradigm

Eben Alexander III, M.D.

The Revelations of Primordial Consciousness

Photo by Ali Johnson Photography

Questions to Ponder

Dr. Alexander doesn't avoid words like God and prayer. How does he define them and why?

How did his experience of an NDE change this neurosurgeon's understanding of brain function and consciousness?

He replaces the prevalent materialist worldview with metaphysical idealism, as influenced by quantum mechanics. Explain.

I was born in Charlotte, North Carolina, in December of 1953. *You're a Sagittarius; does that mean anything to you?* It's beginning to mean more but I don't claim to have much of an understanding of astrology and what it really means. I am respecting it more as a system of understanding and predicting patterns. *I look at it as* *like a karmic map of what we bring in, which can be useful, although there's no explanation that I can think of as why the constellations would impact our personal lives.* If you assume that there are oscillations, predictable back and forth patterns that appear in our individual lives or societies at large, then it makes sense if they align with orbiting timekeepers (planets or moons) in the skies above. It's not that the planets would necessarily be causal as much as being linked to a periodic oscillation, so I am certainly open-minded enough to look at major astrological predictions in that fashion. *Psychiatrist Dr. Mitchell Gibson* said he got interested in it when he saw patterns of people with bipolar or anxiety or depression. When he looked at the astrological influences, he said he could look at someone's chart and say, "You have a disposition to being bipolar," or whatever the issue. Do you know your Myers-Briggs personality type?* I'm an ENFP but I have no real knowledge of what that means.

What were the influences in your childhood that led you to be a physician? My adoptive father was a tremendous influence in my life. His own father had been a general surgeon and took my dad to the Presbyterian Church in Knoxville, Tennessee, every Sunday of his life, back in the teens and 1920s. My dad had a very profound religious belief in a loving personal God and the power of prayer. He got through World War II as a combat surgeon and I think it was his belief in God that brought him back relatively unscathed. He went on to be the head of the neurosurgical training program at Wake Forest University. He was by training a hardcore scientist, yet for him there was never any conflict between his belief in a powerful loving God and his knowledge of science. I grew up in the 60s and the 70s and like many who grew up in that era, I knew that science is absolutely the pathway to truth.

You've spent most of your life in the South. Do you think there are regional influences on you? I spent much time up in New England. I would say I still consider myself a southerner but I thrived in the hardcore academic environment of Harvard Medical School and the very high-end medical endeavors that you run into in that environment to improve the field of neurosurgery. Institutions like that can really open the doors to incredible productivity because of the number of people who are engaged in such lofty pursuits. I thoroughly enjoyed my 15 years teaching at Harvard Medical School but I moved back to Virginia in 2006 after getting a bit fed up with some of the medical politics.

Boston and Harvard are the home of integrative medicine centers and mindfulness meditation with Jon Kabat-Zinn. I would also put in a very strong note for Duke University, where I went to medical school (class of 1980). When I was a student there, we had acupuncturists on staff although they didn't necessarily have an explanation of the mechanism. Duke was also the home of the work of J.B. Rhine who headed up the parapsychological research lab. Duke is a very advanced allopathic medical school that paid attention to successful modalities that weren't explained by Western medicine.

I've encountered many individuals at various places in Europe and the US who are studying integrative medicine, fostering its adoption when benefits are found, as with this new interest in psychedelic drugs. For example, the research in psilocybin for the treatment of fear-of-death in cancer patients and the treatment of alcoholism and addiction, especially very tough ones like opiate and nicotine. Some centers take psilocybin seriously, like New York University Medical Center and Johns Hopkins in Baltimore that have done some ground-breaking work. It's important that we're opening the door to such profound mentally interactive drugs, such as psychedelic drugs, although I do not recommend their use in a casual, recreational setting. I believe these drugs "thin the veil" (traverse the brain's filter) to allow an expression of free will that goes beyond our ego sense and allows for profound healing.

These studies open the door widely to placebo effect, to

the power of mind over matter, which modern medicine has acknowledged for decades in placebo-controlled trials. However, what most people fail to realize is that there are some extraordinary stories about placebo effect where people completely cure their own cancer, infections, even some congenital deformities, and degenerative diseases, through the belief that they are acting in such a fashion to do so. We cover placebo in detail in *Living in a Mindful Universe*, co-written with my life partner Karen Newell, describing the strengthening of placebo effect and the spontaneous regression of advanced diseases that suggest the reality of mind over matter.

We live in a mental universe, as some quantum physicists have come to realize. It's a mistake to think that we are slaves to the physical vagaries of subatomic particles, all simply obeying the laws of physics, chemistry, and biology. The belief that such "bottom-up" causation fully explains all of the events of human lives is just not true. That's why I think that the awakening of humanity--and especially of healthcare and scientists--to the reality of mind-over-matter and the power of consciousness in determining events in our lives is important. This awareness is spearheaded by NDEs and also by phenomena like placebo effect that have been misunderstood in modern medicine.

In Living in a Mindful Universe, *you described these changes as a paradigm shift that is the biggest revolution in science ever.* Yes, although many people would point to the Copernican Revolution as being a major revolution. Copernicus pointed to the sun as the center of the universe at a time when the prevailing view assumed the earth to be at the center. This revolution we are talking about, in which consciousness is fundamental and physical matter only exists as a projection from consciousness, will make the Copernican Revolution look relatively minor by comparison.

If we go ahead in time maybe 50 years, how would you guess that health care will change? We will acknowledge the power of not only individuals to influence their own health through a broadening of their spiritual journey, but also that practitioners will be deeply involved, recognizing that the beliefs of the practitioner also can influence the outcome. Many of our culture's dominant beliefs, inspired by scientific reductive materialism, are falsely restricting. Practices like meditation, centering prayer, and seeking the much bigger picture of why we have an illness or an injury, can assist with hardships in our lives that require healing in a physical, mental, or emotional sense. Ultimately, all such healing is spiritual at its core.

One thing missing in our modern world is the widespread acknowledgment of the spiritual nature of our own being and the spiritual nature of the universe, well-supported by findings in quantum physics (the "mental universe"). We'll come to realize that phenomena like the placebo effect can certainly happen when you feel a higher sense of well-being and reuniting with the higher soul to elevate you over the battleground of your ego, when you realize that the ego and the little repetitive thoughts in your head are not who you really are. We realize that we can only make sense of our existence here when we understand that we've been here many times before and are not limited to this one physical incarnation. This much bigger view of self and a reason for existence can lead us towards a true wholeness, which is what healing really means.

As we realize that challenges like illness and injury are stepping stones or pathways towards higher awakening, we can achieve various physical, mental and emotional healing by acknowledging our profound spiritual connectedness with the universe and the sense of purpose in our lives. I think that is what will revolutionize healing. The most extreme examples of healing occur in NDEs, like when I had an advanced case of bacterial meningoencephalitis, affecting all eight lobes of my brain, that should have killed me. By my seventh day in coma, my doctors estimated a 2% chance of survival and no chance of recovery. How did I have such a profound experience when all the medical data indicated I had no possibility of recovery, and should have had no ability to experience conscious awareness? That's why many in the medical profession are so amazed by my journey to the point where some of them say it can't be true with medical records showing such damage to my neocortex. And yet within eight weeks after waking up, I had a complete recovery and more than complete return of memories that had been totally deleted by the meningitis during my coma. When I first came out of coma, my brain was an absolute wreck.

The medical facts of my case are available in a case report in the September 2018 issue of *The Journal of Nervous and Mental Diseases*.[1] The four-page case report goes through my medical records and finds how astonishing it was that I could have an NDE experience and recover so completely in the face of this kind of illness. It is really mindboggling from a medical viewpoint. If you want to access that medical record, go to ebenalexander.com and you'll find the link to the report.[2] The three independent physicians who wrote that medical report raised the issue that perhaps my extraordinary healing was linked to the extraordinary spiritual nature of the NDE I had deep in coma. That explanation was crucial to the peer-reviewers making sense of the case and accepting it for publication.

Likewise, if you look at the story of Anita Moorjani described in *Dying to Be Me,* with her advanced lymphoma she should have been dead within hours after she went to an ER in Hong Kong in 2006. She had a profound NDE where she knew that her cancer would evaporate, which it did. Or look at Dr. Mary C. Neal who wrote *To Heaven and Back.* She was an orthopedic surgeon who had a profound NDE during a kayaking accident in 1999 when her kayak was wedged under rocks 10 feet underwater. She broke both her legs and was pinned underwater for more than 30 minutes. Western medicine suggests she should not have been able to come back from that, but she had a complete recovery. These extraordinary and shocking examples of healing completely defy Western medical understanding, yet they do occur and involve rich, profound spiritual journeys. The more we come to recognize that we are spiritual beings leading a spiritual existence in a spiritual universe, the more progress we can make. The important takeaway is that all of us can benefit from these kinds of stories and the medical profession can learn much about healing from studying these stories and events.

It seems to me that illness is often a metaphor for some emotional and spiritual issues, like breast cancer is about nurturing one's self. Did you see emotional patterns in patients that you worked with doing brain surgery? Yes, very commonly. I saw that the patients' attitude had everything to do with how they were going to fare. Those who were too despondent, hopeless, and lost due to what the other doctors told them about their limited life expectancy, etc.––if I couldn't get them out of that funk and get them on board with believing that they could get better, it was hopeless. *Did you see specific metaphors like, "I spent too much time in my brain or I am too logical and inhibited my emotional life, so I have to pay attention to brain restructuring?"* Yes, there is certainly evidence and some respected medical studies that various personality types are linked with various cancer types and other diagnosis, but a lot of that research is still preliminary work.

When you were in a coma, your personal memory was erased, but did you have any sense of your own past lives that led you into this lifetime? No, it's important to stress that one of the really unusual features of my experience was the fact that I was completely amnesic about the life of Eben Alexander. I had no language, no personal memory, none of my scientific knowledge. I didn't even recognize loved ones at the bedside when I came back to this world. I only came to realize later how important that amnesia was. In spite of the fact that NDEs have major similarities across cultures and across millennia, they are always tailored for the individual. I was a neurosurgeon who wanted to know more about consciousness and about the nature of healing and that's why I went through such an incredible journey.

I had to be amnesic and it had to do with the beautiful girl on the butterfly wing, a guardian angel who I did not identify when I first came back to this world. She was a deep mystery to me. I had never read the NDE literature before my 2008 coma and her presence was a real shocker. It demanded a deeper explanation which really didn't come to me until about four months after the coma. *That was your biological sister who you didn't know.* Yes, that was the mind-boggling realization I reported in the book, that finally crystallized my acceptance of the reality of the entire experience, and drove me into a lifetime of commitment to explaining it. But I also wondered why my father wasn't there in the NDE. He had passed over four years before my coma. The more I read the NDE literature after my coma, the more I realized that if I had scripted my NDE, I would have put him there front and center and yet he was nowhere to be found. A meditation that occurred two years after my coma (recounted in *Living in a Mindful Universe*) included revelations about why he could not be "apparent" to me during the NDE.

I falsely believed before my coma that memories were stored in the neocortex and so to have my memories completely deleted by a disease that primarily attacks the neocortex would seem to support the belief memories are stored in the brain, which I now know not to be true. *Because they came back.* That understanding of my recall came from specific conversations with close family and friends about early life memories and our earlier conversations. I know now that memories are not stored in the brain, as discussed in detail in *Living in a Mindful Universe.* It's one of the biggest pieces of evidence that the brain is not the producer of consciousness, or the site of memory storage, but is only a filter or reducing valve, a transceiver, that allows us to access information. There will be much more coming in the next years as I refine this understanding and the deep message about consciousness and the nature of reality.

In the midst of the NDE journey I saw in a generic fashion how vast civilizations, far more advanced than ours, go through a similar process of sentient beings incarnating in the physical realm with a temporary kind of dumbing down and forgetting of information. That's why we don't all have the information that's available to our higher souls through our whole life. Scholars who studied past-life memories in children realized that by age six or so many of those memories disappear because there is an active process of memory suppression (or "programmed forgetting," similar to how most of our dreams are treated, even though we know sleeping and dreaming are crucial to health for humans and animals). I saw the life review as interwoven threads in this tapestry of reincarnations, like beautiful demonstrations of flying fish that would be in and out of the water representing our being down here in the murky depths of dense four-dimensional space-time, then being liberated when we're freed from the shackles of the physical brain between lives.

An NDE is an incredible explosion of information to reframe the worldview; that's why NDEs can be so shocking and mine was absolutely that. Here I am 10 years out from waking up from that coma and I'm still deeply mired in the beautiful process of unraveling and coming to some understanding of it. As a neuroscientist, it is very interesting to understand the nature of reality regarding the relationship between brain and mind. My personal story parallels the journey of the modern scientific community and some of the deepest truths that we're approaching through the measurement paradox of quantum physics, the hard problem of consciousness, and the evidence for non-local consciousness in that we can know things beyond the ken of our physical senses.

All of this is driving us to a much more refreshing worldview of humanity, which is why we're all here and where we're headed. The NDE community does a tremendous amount to remind us that at the very core of the universe is a very infinitely loving force; a creative source of all that happens that is intimately related to our conscious awareness. That's where the notion of free will of the higher soul starts to take on tremendous power to bring this world back into alignment with what we came here to do. In many ways we've painted ourselves into a corner, but I think there are many positive signs that we're headed out of that crazy morass, the false sense of separation that occurred with the divorce between scientific and technological capabilities in the 20th and early 21st century and the spiritual nature of human existence. It is time to annul that divorce.

When I read Proof of Heaven, *I thought this is similar to Robert Monroe's experiences, yet I assumed that you hadn't read his books.* Right, I had never read anything about Monroe before my coma; it was only afterwards that I started finding out about him and his work. *Do you agree that his experiences and yours were similar, exploring multidimensions and love?* Yes, very much. I worked at the Monroe Institute daily (2010 to 2011) where all the people he worked with were commonly encountering souls of the departed but if you mentioned spirituality to Monroe, you might end up getting fired. He was way too much of a pragmatic down-to-earth businessman. He had been a successful radio producer in New York City and didn't want to hear about woo-woo spirituality. Even though he was dealing daily with souls of people who had left the physical plane, you weren't supposed to put that in any quasi-religious or spiritual terms. He was all about the engineering. *The right-left brain balancing with sound to astral travel?* Exactly.

In Proof of Heaven: A Neurosurgeon's Journey into the Afterlife *you describe your unexpected and total change of understanding of how the universe works, after being in a coma and having an NDE. Please distill for us how we can apply what you learned to our lives.* Our conscious awareness is NOT created by the physical brain, but it is filtered by the physical brain.

Consciousness is primordial in the universe, the ultimate creative and evolutionary force, providing top-down organization from the mental level of the universe, one of full assimilation and integration of information. Humans have access to this mental layer, which guides the emergence of a unified reality. By avoiding mention of consciousness, some quantum physicists are forced to accept something like Hugh Everett's 1957 Many-Worlds Interpretation of the measurement paradox in quantum physics (infinite parallel universes emerging at every choice point of sentient beings). Objective idealism is a philosophical position that better explains the primacy of consciousness, as we fully argue in our book *Living in a Mindful Universe.*

This dovetails into many of the deepest questions in modern science having to do with the nature of consciousness and the relationship of brain and mind. For example, the measurement paradox in quantum physics has been extraordinarily challenging to understand, partly because we made some assumptions in the scientific revolution over the last 400 years that just weren't true. Until we started to examine very closely brain, mind, and consciousness, we couldn't tell that our underlying assumptions were misleading in that they failed to explain a tremendous amount of human experience. I think that people should rejoice in this extraordinary shift of the tide in our understanding of the nature of reality from a scientific perspective––it's gigantic.

It opens up the possibilities of the afterlife by demonstrating very clearly that the brain is not the creator of consciousness as much as a filter that allows conscious states to manifest. This is where it really gets extraordinary because we can account for NDEs, after-death communications, psychic medium communications with souls of departed loved ones, and so on. We start to get some reasonable way of explaining things like the extraordinary scientific evidence for reincarnation. If you look at the work of Ian Stevenson and Jim Tucker after more than six decades of disciplined scientific work at the University of Virginia (uvadops.org), they've uncovered more than 2,500 cases of past-life memories in children strongly suggestive of reincarnation. To simply deny it as impossible just because our current theoretical models don't explain it is ridiculous, especially when you realize our current explanatory scientific models don't tell us anything about the fundamental nature of consciousness. The world of neuroscience is about as far from any notion of the deep understanding of consciousness as we've ever been.

My NDE in many ways is a very direct refutation of the simplistic and false paradigm of materialism or physicalism that says that the physical world is all that exists. If my NDE were the only one out there, I don't think anybody would pay it much mind. The fact is it's one of millions of NDEs, and when you study them in detail you find they have tremendous similarities, suggesting they refer to a realm more real than this one. It gets very exciting when you realize that current scientific study of consciousness involves various systems, as we explained in our book *Living in a Mindful Universe*. Idealism and the mental universe appears to be the best model to explain much of what happens in reality. In many ways it returns free will into human activity because the materialistic model that I worshipped before my coma, the belief that brain creates consciousness, basically denies that there is any such thing as free will. It pretends that all the workings of consciousness and phenomenal human experience are nothing more than the confusing epiphenomenon of chemical reactions and electron fluxes in the physical substance of the brain, but that is simply not true. That materialist paradigm very quickly stomps on any notion of free will, the notion that humans have anything to do with determining the course of events that occur in this four-dimensional space-time reality.

Free will is alive and well, as we explore deeply in *Living in a Mindful Universe*. It has to do with acknowledging that our choices matter and that we have responsibility, in contrast to materialists who try to pretend it's just chemical reactions and can't see any way to work free will into that equation. Yet we have no responsibility for any choice we make if it's just these chemical reactions following the laws of physics, chemistry, and biology. We ARE responsible for our choices; we WILL reap what we sow. That's why the mindful universe is very important; it's very mindful of every bit of what we do. In fact you really can't separate our kind of conscious awareness from the self-awareness of the universe at large. That's where I believe all of this gets very exciting and liberating for the individual seeker who wants to do some good for this world.

We know that there is more than the little fatty watery brain. What do you think is the mechanism that allows for telepathy, distant healing and all the paranormal experiences? Does quantum physics explain it? Quantum physics definitely points out very clearly that our notion that the physical brain creates consciousness out of physical matter alone is completely false. You see it when you start studying all the evidence for non-local consciousness in the reality of things like telepathy. For example, in identical twins, one twin might burn her finger on a stove and the other twin could be 1,000 miles away but feel the pain and develop a blister.

Daryl Bem's experiments indicate that we all have the power to know the future before it happens, even before the future has been generated if you look at the experiments that use RNGs to select pictures on a computer (horrific, neutral, or pleasant), which is what Bem demonstrated in his experiments. *His experiment was very controversial; some said his—and other social scientists'—methodology was flawed. Bem maintains meta-analysis indicates presentiment does occur.*[3] I've read the criticisms of Bem's methodology and find them lacking, especially in his meta-analysis the results are quite clear, without flaws in method or reasoning. Experiments of nature like NDEs that occur naturally in people may contain significant clues about the deep structure of the universe. Sometimes labeled "anecdotes," they often teach amazing lessons about reality that can't so simply be set up in an experiment.

Dean Radin has also done experiments in presentiment, demonstrating that your autonomic nervous systems (which are the basis of our fight or fight responses) respond to the next picture, even before the RNG on the computer has selected which picture to show. Psychic spying, or remote viewing, has been demonstrated beyond any reasonable doubt by statisticians like Jessica Utts and yet people refuse to look at the evidence, including *The New York Times* science section and *Scientific American*. They are only going to print the studies that they think are mainstream science, which is often leading us away from the underlying truth.

Quantum physics is at the core of it all, because it is the most fundamental presentation of the mind-body question. The place where material science got so far off the rails was assuming that the physical world is independent of the observing mind. Yet quantum physics shows very deeply and profoundly that any assessment of the world by the intelligent observing mind leads to paradoxes that prove that consciousness must be fundamental. The founders of quantum physics, like Paul Dirac, Wolfgang Pauli, Erwin Schrödinger, Max Planck, John von Neumann and Eugene Wigner, all made very profound statements about how consciousness is primordial in the universe, based on quantum physics experiments. The experiments have gotten more and more refined in proving exactly that. *The observer changes the potentiality of what the observer is looking at.*

One of the reasons I think all this has been so controversial and difficult is because, built into the assumptions of 400 years of the scientific revolution, is the notion of what Einstein referred to as local realism. That's the idea that nothing occurs in that four-dimensional space-time without a local interaction of subatomic particles, and that any such observations yield results independently of the observing mind. So, for example, when I look up at the star Arcturus in the night sky, a photon from that star must come through space and interact with the pigment in my eye and it's that local interaction that gives me the information to tell me the star is there.

Quantum entanglement is a mystery, how subatomic particles seem to be linked together so that any measurement of one of them instantly influences the other particle––no matter how far away. This is what disturbed Einstein to no end because in special relativity, which he developed, information can only travel at the speed of light, no faster. But this entanglement implies "spooky action at a distance" in that the information transfer is instantaneous. Einstein thought there must be hidden variables, as elucidated in the "Einstein-Podolsky-Rosen (EPR) Paradox" in a 1935 scientific paper. Then John Bell, the Irish physicist, came up with ways to turn the EPR ideas into experiments and wrote a 1964 paper about that. In the early 1970s, others started designing experiments to assess hidden variables of Einstein's theories versus spooky action at a distance or entanglement, all based on John Bell's inequality, since most of the founding fathers of quantum physics acknowledged the apparent primacy of consciousness.

Every one of those experiments from the early 1970s up to the current era clearly showed the reality of spooky action at a distance and entanglem*ent, showing that Einstein was wrong, there are no hidden variables. The "realism" part of local realism* is the notion that there is a world out there that exists independently of us. And that's where it falls apart because quantum physics strongly suggests the world does not exist independently of the observing mind. It certainly opens the doors to the fundamental nature of consciousness, to the reality of parapsychological phenomenon, to the reality of an afterlife and reincarnation as explored in the incredible work done by Ian Stevenson and Jim Tucker at the University of Virginia.

How would you define consciousness or mind? Is it a lawful neutral universe or do you imply that there is love and helpfulness involved in it? Consciousness is fundamental. Accounts of the NDE go back 2,400 years, all the way back to the time of Plato, when the Armenian soldier Er was killed in battle, but came back to life on a funeral pyre days after he was killed. In this early reported NDE, Er claimed you go through a life review with your life flashing before your eyes. All the pain and evil you've handed out to other people will be experienced during your life review as if you were the others who were influenced by your actions and thoughts. It's one of the most beautiful examples of how we are dreaming the dream of the One Mind. It's a very good reason that the golden rule, "Treat others as you would like to be treated," is written into the very fabric of the universe.

At the deepest levels of the Core, I became one with that mind, with that God force. Of course in the Methodist church I grew up in we were not allowed any kind of talk about things like reincarnation or becoming one with God and yet that's the key message I got from my NDE. When we are freed up from the shackles of the physical brain, we return to that Oneness. We first go through this life review that suggests what we can do in the next incarnation to best learn and teach those lessons with our soul mates and soul group. Then we can access the higher realm, that pure Core of Oneness with the divine that I experienced so fully. The entire physical universe emerges from that realm of mind. Every one of us is sharing that one mind.

A beautiful book that addresses some of this is my good friend Dr. Larry Dossey's* *One Mind*. He wrote it because he is an identical twin and knew that he had felt telepathic experience with his brother [*as did his wife Barbara with her twin*]. He wrote not just about twin studies and evidence of telepathy and clairvoyance, but also about relationships between animals and humankind that show a telepathic connection. It's a profound concept that all this really starts in the realm of the mental and that that's where a lot of our power resides, including the power of the placebo effect and spontaneous regression with advanced diseases, our very source of true free will. That can only occur when you postulate a top-down causality. Before my coma, as a scientist who paid some attention to quantum physics, I thought it was all bottom-up, that somehow we were trying to explain the workings of this universe based on subatomic particles following the laws of physics and the atoms, molecules, the laws of chemistry, and then biology. Top-down causality is discussed in quantum physics, as by South African mathematician George F.R. Ellis, which has everything to do with the One Mind and its filtering into our limited conscious awareness by the brain.

This view is actually quite comfortable for a lot of quantum physicists; for example, Richard Conn Henry (the Head of Astrophysics at Johns Hopkins University) wrote a beautiful one-page essay in the scientific journal *Nature* in 2005 called "The Mental Universe." He clearly points out that from a quantum physical standpoint, we live in a mental universe where causality is in the mental realm, not in the physical realm (hence the crazy idea of the Many-Worlds Interpretation of infinite parallel universes, accepted by many quantum physicists, as the price paid by failing to comprehend the power of objective idealism). This means we have tremendous power over the events in our lives with the help of centering prayer, deep meditation, and differential frequency sound-enhanced brain entrainment like Sacred Acoustics (sacredacoustics.com). I've used those differential frequency sounds for the last eight years to help get back into my NDE, not only to recover the memories, but to develop rich relationships with entities, powers, and the forces of that NDE. The Sacred Acoustics tones influence the superior olivary nucleus complex, a circuit in the lower brain stem that is more than 300 million years old. Most music and sound we hear is primarily processed in the recently-evolved (last 1 to 10 million years) acoustic cortex. By affecting that timing circuit in the lower brain stem, Sacred Acoustics tones allow differential frequency brain entrainment to modulate ascending signals normally crucial in binding consciousness to engender rich transcendental journeys into conscious awareness and contact with primordial mind across the veil.

You talk about the inner Core, is that equivalent to God? God is all through every bit of this: You can't leave God out of any part of it. That infinite sense of living oneness, well-being, comfort, trust and a sureness that I felt in the Core realm was absolutely bathing in the ocean of love of that divine creative source of all that is. The God force is identical with our sense of conscious awareness, but our little ego and little linguistic brain can lead us so far astray with the trickle of awareness associated with our normal waking consciousness. That's the great wisdom that can come through deep meditation and centering prayer, by "traversing the veil" of the filter of the brain to acknowledge the oneness of connection, the sense of purpose and having true free will, not just the reflexive automatic responses of our ego self--that can greatly influence how events in our lives unfold. I believe this allows more ready interpretation of the course of human history as representing a true free will, and not just a predictable pattern of evolution due to the vague chaotic swarming of subatomic particle following the laws of physics.

God was apparent to me as a beautiful breath of fresh air, a perfect soft summer breeze, or the divine wind I felt when I was on the butterfly wing in the Gateway Valley when I was in coma. That God force of pure love was also apparent to me in the prayers of all those thousands of beings surrounding me in the distance at the end of my coma journey. The part of this belief that defies a lot of our conventional orthodox Christian thinking is becoming one with God in spite of the fact that Christ said as much when he said that the kingdom of God is within you. Christ was pointing out that each and every one of us has that God force as part of our very being. You know it through Gnosis, the personal experience of going within, so meditation, centering prayer, and differential frequency enhancement of meditation are tools that can all open us up to this greater realization of who we are and why we are here.

I am far more of a scientist now than I've been before and realize that the materialist model is fatally flawed. It should have died off with the advent of quantum physics and yet it has held on with extreme tenacity in spite of evidence to the contrary. But this is why I think books like yours and those of the other thought leaders in this book are so important because they connect the dots in a scientific fashion to the reality of our spiritual nature. There's no way out for science or spirituality as long as they continue to profess this divorce. As long as they keep sticking to materialist models and trying to pretend that there is no such thing as consciousness, which is fundamental as an organizing top-down principle in the universe, they are going to be forever frustrated circling the deep mystery of the measurement paradox in quantum physics. By putting consciousness first, by looking at the brain as a filter that allows in primordial consciousness, it all starts to make far more sense. Then we can move forward with models that actually make sense and help us explain the realities of not only the afterlife but of reincarnation. Every bit of that fits perfectly in a world that accepts consciousness as fundamental; the deep lessons of quantum physics are teaching us exactly that.

Do you think it's typical for human nature that when a paradigm shifts there's tremendous opposition and you get put in jail or burned at the stake or whatever? Thomas Kuhn made important points very applicable to this paradigm shift we are seeing now, just as we spent the last 80-plus years circling the drain trying to understand the measurement paradox in quantum physics. Of course, Max Planck is famous for saying that the way these scientific revolutions occur is one funeral at a time. I'm not quite that pessimistic because all you have to do is pay attention to the evidence and follow rational argument concerning it. For the serious scientific mind desiring more along these lines of inquiry, I recommend the books by Edward Kelly from the University of Virginia Division of Perceptual Studies: *Irreducible Mind: Towards a Psychology for the 21st Century* and *Beyond Physicalism: Towards Reconciliation of Science and Spirituality*. The latter is an incredibly powerful book showing the way this revolution is going beyond physicalism toward the reconciliation of science and spirituality.

When you start running into situations like most physicists explaining the measurement paradox by saying all you need is infinite parallel universes and Hugh Everett's 1957 many-worlds interpretation of quantum physics, then they never have to discuss "consciousness" but that is a tough worldview to work into one's daily life! A different way of solving that problem is idealism, realizing that mind is primarily responsible for everything that happens in this universe and it returns free will right where it belongs. Mind over matter is not a new concept. In medicine for more than 60 years we've looked at the placebo effect. Go to the Noetic Sciences website (https://noetic.org) and look at the book they put together in 1993 called *Spontaneous Remission: An Annotated Bibliography* on spontaneous regression and healing. You'll find more than 3,500 cases of extraordinary healing often because patients just believe they can get better, including spontaneous regression of tumors and infections. It's a beautiful example of mind over matter.

Placebo effect is a rich example of mind over matter. If you don't believe it's real, ask Big Pharma; they realize they've got to overcome about a 30% beneficial effect that happens simply by patients believing they can get better. This phenomenon shows that the materialist science and molecular medicine taught in medical schools is quite outdated. *My favorite example is people who have Dissociative Identity Disorder, one alter will have diabetes and the other won't, one will wear glasses and the other not.* Interesting that you point out that whole dissociation phenomenon because this understanding was first made (to my knowledge) by Jon

Klimo* who is renowned in the UFO world. He is a renowned psychologist and parapsychologist who has provided a sound argument for dissociation to explain much of the distribution of consciousness. Bernardo Kastrup* has recently written some very illustrative articles concerning dissociation and how it might explain much of this world we inhabit. There is a whole top-down ordering of causality in this universe that has to do with the mental qualities of the universe.

Did you do some kind of centering before a surgery or did you focus on the left brain, I've got to do this and this. I would use prayer, which I believed for much of my career, although eight years before my coma, I became agnostic. I lost my faith in prayer and a loving personal God. That had to do with a phone call from the social worker and a perceived rejection from my birth mother. But, for much of my career, I believed in prayer and I saw the operating room as a very peaceful, harmonious place. The nurses and residents would take all the calls, so you could focus on what you were there to do. For me as a neurosurgeon, it meant a lot to be right there with my hands, trying to help patients to get to their very best. I dealt with a lot of malignant brain tumors, tough aneurysms, vascular malformations, many cases of severe chronic pain and conditions that required deep brain stimulation but I loved every bit of it because often I did help many of my patients get better. Really, it's a very humbling experience.

When I was in college, I was on the sport parachuting ("skydiving") team at the University of North Carolina-Chapel Hill. We put together an eight-man star that no other college team at that time could accomplish. Our competition involved free-fall star-formations where, for example, you jump out of an airplane at 10,500 feet. You've got 45 seconds to build yourself into formation with these other 12 people flying in free fall. We competed better when we meditated using Silva Mind Control to practice our maneuvers mentally before the jump. I was not clear on how it might help with other life activities, so I left meditation behind after college. Now of course I am a huge proponent of meditation for all purposes, for creativity, for guidance, for improving health, gaining insight and connection, and strengthening the immune system: I can't imagine anyone today trying to live without meditation.

You and Karen do meditation playshops around the world. What's the focus of a playshop? We don't like the word "workshop" because there's nothing about this that really looks like work. I try to meditate an hour or two a day and I look forward to it. For me it's a beautiful way to engage my higher soul and my free will to try and manifest the world that I want to see. There are many reasons that people come to these playshops, but it's really all about healing and becoming the soul that you came here to be. We use differential frequency sounds as I said earlier, the sacred acoustics. I would also encourage people to visit my website (ebenalexander.com) to find the free online course, 33-day Journey into the Heart of Consciousness, that covers many of these deep concepts. More than 8,000 people have taken that e-mail course. There is a translate button, so we have people from all over the world participating who leave many stories and helpful comments for others.

I think readers would like to know when the most difficult times in your life have been and how you coped with them? One was the perceived rejection by my birth mother in February 2000: I tell that story in *Proof of Heaven*. There were some challenges around the neurosurgical work, mainly having to do with the medical politics in some of the institutions where I worked. Even though they were very renowned institutions, they struggled for more research money, which created some conflict within the institutions. Even though I had done very well as a physician at those hospitals, that made me a bit of a target. My publication list (http://ebenalexander.com/about/publications/) lists more than 100 peer-reviewed papers in the medical literature and more than 50 chapters and invited articles, all about trying to improve the practice of neurosurgery. I found it a shame that the financial battles within the departments where I worked could be so devastating. It was sad to see that much disruption of systems that were trying to provide good medical knowledge taken down by the vagaries of medical politics. Other than that, I've been very blessed and had a wonderful family life.

What about your divorce after your first marriage? Some people say divorce is a graduation. My former spouse and I had some marital difficulties for ten years before my coma. I thought that the coma experience might help to mend our marriage but we both came to realize that we were on separate pathways. I am still very close to my former spouse and in communication almost daily. We obviously have a big interest in raising our two sons who are now 20 and 31 years old. Divorce occurs in about 80% of cases after an NDE, even if you had stability in that marriage before. An NDE is such a profound life changer that sometimes it just puts you in an entirely new orbit and a prior marriage doesn't survive it.

Are your two sons on your same wavelength in terms of thinking about spirituality? They are both wide open to the realities of meditation, of telepathy and they talk about afterlife and reincarnation. Both of them are fans of Sacred Acoustics. Eben, my older son, would listen to Whole Theta while studying and his improvement on medical board exams went up significantly. He listened to it for more than 400 hours while he was studying. My younger son, Bond, has found Sacred Acoustics tones useful for lucid dreaming and OBEs.

You mentioned that there is material that's yet to come based on your NDE. What's your next book about and what are you exploring now? The next big step is to try and connect the dots more fully. *Living in a Mindful Universe* attempts to bring science and spirituality together and to explain consciousness in ways that can make sense to all of us. I am trying to refine that, especially building up the bridge from the scientific side into the quantum physics side. How is the brain doing this? I realize how every neuron can serve as a quantum computer because every neuron is basically the playground of Heisenberg's uncertainty principle. You must remember that none of that world out there occurs independently of you, every bit of it is a projection from mind. Even though the world does follow apparent laws of physics, chemistry, and biology, there is a principle of sufficient reason and the top-down causal principles discussed above.

The bottom line is that the main events in human lives are all driven by top-down causality which begins with the mental universe. Quantum physicists admit consciousness is the primordial nature of the universe that gives us the physical universe coming out of it all. The fact that it does work this way is seen in the placebo effect and the spiritual healing found in NDEs. You can't explain that through any materialist model at all. Already we've got experiments of nature set up to prove the reality of this kind of understanding. The scientific experiment of double-blinded randomized controlled trials is not adequate for all modes of knowledge. In fact, as theoretical physicist Richard Feynman said in his Nobel Prize acceptance speech, there is plenty of information that will come to be known that cannot be proven through scientific methodology. Nikola Tesla said, "When science begins to investigate non-physical phenomena, they will make more progress in the first decade than in the entire history of science." This is really about each and every one of us starting to experiment and explore our own consciousness and realize how much power we have over our emerging reality.

Could you comment on the quantum measurement problem and the hard problem of consciousness? Yes and I also want to define another word: *consilience*. This is a term used in science, history, and philosophy. Consilience simply means when you look at a deep profound problem from different perspectives, you find that radically different ways of looking at it tend to align––that is consilience. [*The theme of the Society for Scientific Exploration 2019 conference.*] We have a very profound form of consilience now with two of the deepest profound mysteries known to modern science. One of them is the measurement paradox in quantum physics in what is called *contextuality*. You can demonstrate very clearly in quantum experiments that the mind of the observer and conscious choices of that mind have a tremendous kind of black and white impact on how the experiment evolves and the kind of information that comes out of it. This is the point that John Wheeler, who is one of the most renowned quantum physicists of the 20th century, the head of physics at Princeton University, came up with in what he called "the participatory anthropic principle."

Wheeler stated, "All things physical are information-theoretic in origin and that this is a participatory universe," but oddly, he called parapsychology a pseudoscience. In reality, it has become clear that many of the clues that might have helped Wheeler more fully develop his principles would have resulted from deeper understanding of the empirical findings of parapsychology. We participate in the evolving reality to the point where one of his thought experiments concerns a photon coming from a very distant galaxy and using intervening galaxies as a gravitational lens. This photon that left its place billions of years ago, its very wave or particle presentation, is affected all the way back to its origin, based on the mindful decision of an astronomer to observe that photon.

The founding fathers of quantum physics argued that consciousness is fundamental because that's what their experiments suggested to them. We're finding now that really the only way out is to look at idealism as we explained in *Living in a Mindful Universe*. Metaphysical idealism maintains that the universe is mental primarily, with top-down causation of this mental universe. In the world of quantum physics you're finding the primacy of consciousness written into the very fabric of the subatomic structure of the material world around us. If you go into the hard problem of consciousness, as discussed rigorously by the Australian philosopher David Chalmers in his beautiful book *The Conscious Mind* in 1996, he points out a set of explanatory gaps in postulating that the physical brain gives rise to the conscious mind. The work of John Searle, Thomas Nagel and other philosophers interested in the mind/body question were hot on this trail in realizing you couldn't explain it all from materialism.

In fact, Wilder Penfield, one of the most renowned neurosurgeons of the 20th century, wrote a book in 1975 called *The Mystery of the Mind* based on his intensive scientific investigation with electrical stimulation of the brain in awake patients. Several decades of rigorous scientific work proved to him that the brain can't be the producer of consciousness or of free will. He thought for a long time that the brain was the site of memory storage but the more he tried to find it, the more he realized the brain's overall capacities were insufficient to explain phenomenal consciousness. Around 1975, he said maybe memory might be stored somewhere in the brain stem but not in the neocortex, and he was quite convinced of that. Modern evidence confirms very clearly what he surmised: memory is not stored in the brain at all. That is a death nail in the coffin of the materialist model in neuroscience that people don't talk about. There has never been a case where we resected some part of the neocortex or some other part of the brain and actually had a definable loss of memory.

It's true that if you damage the medial temporal lobe, hippocampus, or entorhinal cortex and structures like that you can interfere with conversion of short-term to long-term memory, but that is not the same as getting rid of the locus of memory storage. Memory is not stored in the brain and this is something we cover in detail in *Living in a Mindful Universe*. Not only is consciousness not created by the brain but memories are not stored there—the brain allows access to memory and experience, but is not the ultimate source. Of course when you realize the scientific evidence for reincarnation with 2,500 cases, and in 35% of those cases the children actually have a physical birthmark that corresponds with the lethal wound of a prior incarnation––that body of evidence is extremely powerful. The only way to come up with the big picture of what this all means about the nature of reality and understanding consciousness is to look at all manner of non-local consciousness, which includes things like past-life memories in children indicative of reincarnation. Once you realize reincarnation is part of our existence, obviously you're no longer pretending that memories are stored in the physical brain.

What do you do for fun being such a productive person? Karen and I love gardening, so in the warm weather we are out in our garden. We have two dogs, we love playing with them and taking them for walks. We have an aquarium with salt-water fish, and we have a pond with 18 koi fish. We have a lot of bird feeders and love the birds that come trouping through the lovely mountains of Virginia. We love the local flora and fauna, so we engage them all we can. I really love time with friends and family, with sisters, with my mother who lived to the age of 99 with a mind that remained quite sharp. I love time with my sons, hiking, camping, skiing, and scuba diving. I enjoy meeting new people, especially spiritually-inclined seekers, and that's why Karen and I love giving these meditation playshops. I never could have imagined any of this, ten years ago before my coma, and it's all a tremendous gift.

I love meditating because going within and exploring the power of mind over matter is an extraordinary adventure. I would encourage every single sentient being reading this to do exactly that: Go within, which is actually going out into the universe, when one fully comprehends filter theory and its implications. Once you realize that going within, mind gains tremendous power, plus it's very entertaining. It's amazing how much knowledge we come to glean by going within. It was a glorious gift spending a week in coma due to a severe brain infection where my neocortex was turned to pus. It shows me how trusting in the universe, realizing that we have a higher purpose and living it fully from the heart is really the best any of us can do. I am very grateful to my family and friends. Karen is my life partner in this journey and I'm very grateful to my former spouse, Holley, who has been a tremendous boost all along including those years since we split up.

All the recent studies show that climate change means we don't have much time before we really radically alter the environment-- maybe a decade. Are you optimistic? The more we keep burning fossil fuels, the more we are going to be driving species to extinction that have taken tens to hundreds of millions of years to evolve. That is a crime against nature. I can't believe they haven't already outlawed fossil fuels in ground vehicles. I mean it's simple enough to make ground vehicles electric and to use all sustainable energy sources available to make them work. The dollar cost of inaction will become prohibitive and has already greatly exceeded the costs of accommodating by ridding ourselves of fossil fuels. Our environment will crumble with disastrous consequences for the entire planetary ecosystem, if we don't start to address climate change and pollution now.

With the rise of autocrats around the world and populist nationalist movements, do you feel like that's a release of the old way, bringing it up to the light to clean it out? I am optimistic mainly because what I see is the deepest lesson coming from NDEs, and the awakening of consciousness acknowledges that we are sharing one mind, that we are all in this together. We should practice the Golden Rule with every breath we take, treat others with love, compassion, respect, and forgiveness when necessary to change this world. We are all in this together; the deepest lesson from

NDEs and from consciousness studies is that we need to treat each other with kindness and respect, we need to take care of the least, the last and the lost.

Reincarnation is real, afterlife is real, life reviews are very real, so we'd be damned fools to keep treating each other poorly. People will start thinking, "If this is the way the universe works, I am a fool to keep treating my neighbor and others with such disrespect and keep handing out pain and suffering to others. I am going to feel the full brunt of that in my life review and that means in my next life the karmic setup is that I have to learn the lesson the hard way, by being on the receiving end of it all." Let's realize that the Golden Rule is real: Treat others (including your*self*) with respect. This is what the scientific study of consciousness reveals as a deep truth of this universe and that is what will change the world.

What's your next book going to be about? There is no question that we still have a tremendous amount of work to do in understanding how ontological or metaphysical idealism works. How is it that we can align with that mind of the universe, which is actually the source of our very conscious awareness? How can we reduce this kind of veiling and filtering between us and much more clearly see the information that's presented to us from those realms and also act on it? How can we use that free will of the higher soul to cause changes in this world that will bring the peace and harmony and prosperity that I believe is promised by this extraordinary vision as we awaken? It definitely involves acknowledgment of the One Mind and also is an acknowledgment that you shouldn't have a tiny minority of people with the power to control everybody else.

Books

Proof of Heaven: A Neurosurgeon's Journey into the Afterlife, 2012

The Map of Heaven: How Science, Religion and Ordinary People Are Proving the Afterlife, with Ptolemy Tompkins, 2014

Living in a Mindful Universe: A Neurosurgeon's Journey into the Heart of Consciousness, with Karen Newell, 2017

Audio CD: *Seeking Heaven: Sound Journeys Into the Beyond*, with Karen Newell, 2013

http://ebenalexander.com/

http://ebenalexander.com/resources/reading-list/

Endnotes

1 https://med.virginia.edu/perceptual-studies/wp-content/uploads/sites/360/2018/09/Greyson_-Alexander-JNMD-2018.pdf

2 http://ebenalexander.com/independent-medical-review-validates-facts/

3 https://www.ncbi.nlm.nih.gov/pmc/articles/PMC4706048/

Bernardo Kastrup, Ph.D.

Contemporary Idealism

Photo by Pedro Henrique Casarin

Questions to Ponder

What is the basis for contemporary idealism versus the prevailing materialism in describing how the universe works?

What is the "hard problem of consciousness" and how does idealism solve the problem?

Why is Dissociative Identity Disorder (formerly known as multiple personalities) a useful analogy for reality?

What are the negative outcomes of materialism and what psychological gains does it provide scientists?

I was born in Rio de Janeiro in Brazil in the early 70s, a Libra. I did the Myers-Briggs test many years ago when I was an INTP, but this changes over time and I certainly have changed a lot over the years.

Your first Ph.D. was in computer engineering and the second in philosophy of mind. What led you to do this philosophical work? I worked with artificial intelligence many years ago where you're always wondering about consciousness because we could see multiple avenues for emulating or replicating the functions of mind. We thought we could, at least in principle, replicate most cognitive functions. But why would these functions be accompanied by experience? Why would there be anything it is like to perform those functions from a first-person perspective? What would make a circuit of opening and closing switches not only process data in a certain way, but also feel something?

I could never wrap my mind around that until I realized the problem was related to an unexamined assumption I was making. The assumption was that something in the configuration of that circuitry would generate experience and without that something experience would not exist. Once I became aware of the hard problem of consciousness, I very quickly realized that it was an artificial problem and that the circuits I was building were just the extrinsic appearance of certain cognitive processes in mind at large, in the universal consciousness. *They* were in consciousness, not consciousness in them. That is what led me to philosophy. Although I have been studying philosophy since I could read, that is what really heightened my interest and led eventually to the series of books I've written. I do philosophy informed by science. Technically speaking, what I do is called metaphysics or ontology. But the term "metaphysics" has come to mean, in the popular culture, something supernatural, paranormal, which has nothing to do with the technical meaning of the word. Strictly speaking, I don't do science; I do metaphysics, which provides a paradigm for doing science.

How did your family get from Brazil to the Netherlands? My family is Portuguese-Danish. Kastrup is a Danish name. I left the country to complete my education in France and Switzerland. From Switzerland I came to the Netherlands as a young man and I've lived here most of my life. All in all, it's seven or eight countries that I have some direct connection with, but the Netherlands is my home. *You speak Portuguese, English, French, German, and Dutch?* I speak three languages fluently enough to have an intelligent conversation or write—Portuguese, English, and Dutch. The others I know only enough to survive.

The focus of your book, The Idea of the World, *(like this one) is examining a new paradigm in science, which some people have said is like the third Copernican Revolution. Do you agree there is a new paradigm in science defined as being idealistic rather than materialistic?* Materialism vs. idealism vs. panpsychism vs. dualism, all these -isms are philosophical positions. In principle, science can be done in a way that is entirely agnostic of any ontological interpretation of reality. Science is about how nature behaves, about the regularities of nature's behavior as far as we can observe. If we set up conditions in this way and you run a system, then that happens—this is behavior. The method of science itself has nothing to say about what things intrinsically are; that transcends the method of science. It is not a method that was created to assess what things are in and by themselves.

Hence, the scientists' saying, "Shut up and calculate." Yes, the climax of this aspect of science has been encoded in that phrase, but science is about behavior and, of course, you can infer much about what things are based on their behavior, but you can't really make an assertion about what they actually are. Science is a relative method about the behavior of one thing in respect to another. The problem is that science has become very identified with the metaphysics of materialism to the point that people conflate science with materialism, as if materialism were the essence of science. It's not. You can do science under materialism, idealism, panpsychism, under pan-computationalism and all the other approaches to the essential nature of reality. Then the question is, "Do I recognize that, at least in some corners of science, there is a change in the metaphysical assumptions that scientists make?" Clearly, a lot of scientists are open to non-materialist views. Whether this is a new phenomenon or whether this has been going on for a long time and people were just in the closet and not talking about it, and now with social media and the Internet it is more visible, is an open question. I tend to believe there has been an acceleration in the acceptance of non-materialist views. Now people think more carefully about the shortcomings of materialism, what is difficult about it, and therefore are more open to alternatives. The jury is still out.

The notion of idealism, that mind is primary and that matter follows from mind, has been around since 480 BCE in Greece. Hindus talk about Atman, our inner core that is one with Brahman and will evolve into this oneness that you talk about in some of your books. Tibetan Buddhist Longchenpa in the 1300s said that consciousness exists outside of space and time. There is nothing new about this view. Absolutely nothing. Idealism, as a metaphysics, has been around for a long time. It is a matter of how well articulated it is and how substantiated it is on an empirical basis. If you're an idealist, you should be able to explain obvious facts of reality in terms of an ontology which postulates that there is only one consciousness in the universe, including the fact that I can't read your thoughts, at least ordinarily, and you can't read mine. Moreover, we all seem to inhabit the same universe, the same world outside our personal or individual mentation. We can't change this universe by a mere act of volition the way we can change the imagery of our imagination. So, we have to explain all that.

Descartes separated spirit and matter in the 17th century. When did science define itself as materialist? Descartes tried to bring peace between science and the church by carving out the world between mind and matter. Mind or spirit was left for the church, so it would leave us alone and not burn us at the stake anymore if we focus on the material part of this dualism. There have been many motivations for this divide, many of which are not based on good philosophical or scientific arguments. They were more socio-political at the time.

When do you trace the beginnings of this current wave of idealism that you're surfing? Idealism was very much invoked at the end of the 18th and beginning of the 19th century with German Idealism. Names like Hegel come to mind and Schopenhauer come to mind. He certainly was an idealist but was not a panpsychist [*everything has consciousness*]. I'm writing a book making this case about Schopenhauer. Then, there came the wave of materialists: Darwin toward the end of the 19th century, Nietzsche proclaiming that God was dead, and philosophy going from there to logical positivism—the idea that all you can really know is what you apprehend through your senses. We got into existentialism, which believed whatever meaning there is in nature is projected by us and not really intrinsic to nature out there.

Science has been so successful in enabling technology and so conflated with the metaphysics of materialism that the widespread adoption of high technology was misperceived as the success of materialism. But perhaps because of that, people began to think more carefully about materialism and its intrinsic ability to explain the qualities of experience—what it feels like to see the color red, what it feels like to fall in love, to be disappointed, what it feels like to have a bellyache. People realized that whichever way you twist and turn materialism, fundamentally you can't explain the qualities of experience. It is at least incomplete, most likely wrong and discardable, and that is a realization of the 1990s and of the beginning of the 21st century. It's been articulated, explicated, and exfoliated over the last two decades or so.

You say that materialism is bogus and you have the freedom to discuss these issues because you're not in academia. What is it about academia that is so straitjacketing? When you are an academic, you are evaluated every year for your performance, based on the papers you publish and how you are judged by your peers. From that perspective, you have to fit into the current value system in order to have a career and have some degree of success. My day job is corporate strategy in the high-tech sector, where I am judged by how much money I make for the company. My employers don't care what my philosophical and scientific positions are insofar as they don't relate to my work. I write what I think is correct. I also publish a lot and in some significant journals.

People are talking about your Scientific American *articles.* I'm glad about that and it's a great magazine, but what I meant was academic journals where you go through peer review. *Scientific American* is a science popularization magazine, the only really good one today. But yes, a part of me is surprised that *Scientific American* has published so much of my work recently, because what I submit to them is very explicitly anti-materialism. It's just well-articulated, I hope, and well substantiated. *Scientific American* proved that its editorial crew is honest, open-minded and truly evaluate articles based on the quality of articulation and substantiation rather than some preconceptions. *Brian Josephson* who, as you know, won the Nobel Prize in physics, has trouble getting articles published in peer-reviewed magazines.*

You say that some of the old materialist assumptions are making us collectively mad and we are in a collective trance—how so? The life we live is a consequence of our belief system. What we experience of the world is not only a function of the world as it is, it's a function of what we project onto the world based on our belief system, not only the experience *per se*. If we adopt a materialist metaphysics and think that all there is is unconscious matter and that consciousness is an ephemeral little spark you get by chance when matter arranges itself in a certain way—that is, as a living brain—then this will color our experience of life. If we believe that once the configuration of matter we call a brain loses its structure or coherence—that is, when we die—then consciousness disappears, this will color how we live life because then we believe that death is the end of our consciousness. We have nothing to lose whatever happens to the world a hundred years from now, for by then there will be nothing left of us here. Let the whole thing blow up. The only possible, conceivable meaning of life is to accumulate material goods because matter is all that exists.

Our emotions become secondary, mere shadows of brain activity that have a survival purpose. If we happen to feel bad, we repress our emotions because they have no intrinsic meaning since they're an accident of a program encoded by evolution to favor survival and reproduction. We never live an examined life because we are always dis-integrated, so to say. Our emotions are buried away because we don't care about them. Instead of having an examined life where we try to integrate every aspect of ourselves—our emotions, our thoughts, our aspirations, our sense of purpose and meaning—by having materialism as the mainstream storyline about what we are and about our role in life, we become collectively mad. We become disconnected from our own nature. After all, what are we if not consciousness?

Materialism says that consciousness is just an epiphenomenon, a secondary effect, a consequence of something else that is much more fundamental. What kind of life can we live under this belief system? It is certainly a life of madness in which we are disconnected from our own nature, from the ground of our being which is consciousness, which is to experience the qualities of life in the form of our emotions and perceptions.

The question is, to what extent is our metacognitive behavior a function of our cultural belief system or of our biological programming? Our need to have sex is certainly biologically programmed, but the choices we make about how and what to consume are to some extent cultural artifacts. They are derived from a belief system which we inherit from the culture, a materialism that gives us permission to not care about the distant future or the generations to come. Even people who do care all the way to their grandchildren—I don't know for sure because I don't have children myself—seem to rationalize their destructive behavior by saying, "I need to accumulate wealth so I can leave something for my children and my grandchildren so they are successful in life." But that is optimizing on a local scale and wrecking everything at the global scale.

You're suggesting we need a new language to understand reality? It could be art or poetry. We could look at seemingly bizarre phenomena like ET abductions or telepathy or remote viewing to see what we could learn from them and acknowledge that personal experience has significance. That seems to be a contribution that you're making. I suspect very strongly that nothing I am saying is really original, but it's a contribution nonetheless, in the sense that it is now articulated in a different way. I think logic is very important. I think that if we get rid of logic then it's pandemonium. Logic is important for the predictive models of science that enable technology.

Although logic is so useful and seems to work for practical purposes, that doesn't mean that it reflects what reality is at an essential, fundamental level. Logic may be a very effective empirical window with many practical applications but to translate what works into what is true is a big step. Moreover, logicians have a variety of different logics. They adopt a variety of different sets of axioms, seemingly self-evident truths on which to base their thinking without clarity about which logic is correct. Maybe none of the choices is fundamentally correct. Why would nature obey the axioms? For instance, "If a equals b, and b equals c, then it follows that a equals c," or "If something is true, it cannot be false. If something is false, it cannot be true." These are the axioms of logic and they come from our innate way of thinking. But you cannot use logic to prove that logic is correct, without circular reasoning.

For all we know, nature doesn't follow logic at its most fundamental levels. It just so happens that logic happens to work most of the time, if not always, in our human sphere. But there can be other logics that are equally consistent with nature, intuitionistic logic, for instance. We shouldn't become too focused on one particular logic, Aristotelian logic, just because it is the cultural belief of our civilization. Nature itself, although it seems to comply with Aristotelian logic—perhaps because that is what we expect it to do—may be fundamentally absurd. Graham Priest has been making this case for a while now.[1] What we see of nature may be just an aspect of it that complies with our expectations because our cognitive system perceives nature in a certain way that complies with the rules built into our cognitive system. I think we should be more open-minded when it comes to the more fundamental questions of what nature is essentially. *The quantum physicists of the early 20th century said it's weird and Einstein said it's spooky. Non-locality, quantum tunneling, and existing in various states at once seem illogical.* There has been a lot of academic work on whether logic is empirical or not. If it is empirical, what does quantum mechanics have to say about logic from an empirical perspective?

An example of how you can use art or metaphor to understand reality are people who were abducted by ETs. How do we use that to make sense of what seems spooky? The so-called "alien abduction phenomenon" is an instance of what people call a "high-strangeness" experience. I don't think alien abductions are actually true in the sense that metal spaceships from another star system came to this planet and physically lifted somebody into the sky and performed medical experiments. If this were the case, our logic would still survive, our physics would still survive, our metaphysics would still survive. However, I think what these high-strangeness experiences suggest is an underlying layer of reality that doesn't necessarily comply with logic. Nature at its deeper layers, below the crust, has many more degrees of freedom than our logical intuitions would dare grant it. In certain non-ordinary states of consciousness, people can cognize those deeper layers that do not comply with logic and physics as we understand them today.

Why was Harvard researcher John Mack, who studied abductees, convinced they weren't hallucinating? Mack died 15 years ago. He used to talk about "ontological shock" as a description of his patients' experiences, which is when you're confronted with experiences that contradict your belief system about metaphysics. These people gain access to an aspect of reality that is ordinarily hidden from us, because of the way our cognitive system filters and weaves our perceptions together to enable our survival. That deeper layer does not comply with the axioms of logic or the metaphysics of materialism. As an idealist, I think all the layers of this onion that we call reality are essentially mental layers. I think when people access, through high-strangeness experiences, a layer of reality that is absurd from our perspective in our ordinary mode of consciousness, they are accessing a hidden layer of mental activity.

Is it just their imagination? It is mental, but not necessarily only their personal imagination. When I say that everything is mental, I don't mean that it is only YOUR mentation or only MY mentation. *Like the Jungian collective unconscious and the archetypes we share?* I don't like the word "unconscious," but yes, I'm referring to transpersonal mentation that correspond to the Jungian idea of the collective unconscious but just like our ordinary experience is mental too. Their experience is certainly transpersonal, but mental nonetheless. I think that is what John Mack was getting to, but unfortunately, he left us before he could more fully articulate it. *Are you saying that those people are tapping into some kind of archetype of ETs?* I'm not fond of literal explanations. I think everything we experience is fundamentally symbolic, including our physical world of everyday life. I don't think that there are literally aliens from another star system at work here, rather a symbolic system trying to evoke something much deeper than any literal narrative could grasp or capture.

Is archetype a fair word? Let's say, yes, they are seeing the image of an archetype, which, according to Jung, is a template according to which mind manifests itself. From that perspective, a symbolic system is based on archetypes. All manifestations of mind follow its basic templates of behavior. Mind is not a chaotic, arbitrary system with random behavior because its dispositions and regularities are templates of manifestation called archetypes. So yes, from that perspective, theirs is symbolic experience based on archetypes.

This makes me think of how RNGs get more coherent when the world mind is focused on an event like the World Trade Center tower bombings or New Year's celebrations. That seems to indicate that even inanimate objects respond to thought. I don't think inanimate objects exist as such. I think the partitioning of the inanimate world into separate objects is something we do for our own convenience. We impose that division on the inanimate world. What is the difference between a mountain and a rock on that mountain? Isn't the rock just part of the mountain? But if it detaches and rolls down the slope of the mountain, now is it a different object? Is the handle intrinsically part of the cup? We divide the world into separate objects purely nominally in an arbitrary partitioning based on convenience. So, I don't think inanimate objects exist to begin with, it's just what we project onto the world. But I do think the inanimate world as a whole is the extrinsic appearance of conscious inner life, like the way your brain and its activity are the extrinsic appearance of your conscious inner life. The inanimate universe as a whole is the extrinsic appearance, the image of transpersonal conscious inner life, the inner life of Brahman, if you want to use Hindu terminology.

Does that mean the RNG has consciousness? Consciousness sounds like it equals mind. I think the inanimate universe as a whole is the image of conscious activity and is what transpersonal conscious activity looks like from our perspective. I don't think it HAS consciousness in the sense of being an entity that possesses consciousness as one of its properties. It *is* consciousness—the extrinsic appearance of consciousness.

Is consciousness more than mind? In Eastern philosophy, people tend to translate the words differently. Mind has more to do with cognitive activity, with thoughts, where there is a subject separate from its object. The subject acquires knowledge of its object, so that equates mind to cognition, to thought, to apprehension of knowledge. In the Western tradition, however, the terms mind, spirit, and consciousness have been used for a long time interchangeably. I tend to use mind and consciousness interchangeably, except in psychology papers where it is necessary to make a distinction. Consciousness does not necessarily entail cognition or a split between object and subject. If you're talking about pure consciousness, you're talking about what it is like to be that thing/entity. Then, the entity is conscious. You can be conscious in that simple way without any conceptualization of objects. I think underlying nature is consciousness, the existence of qualities of experience. I think cognition and metacognition are the product of life, of living beings, not of the inanimate universe.

It seems that you agree with Jung that the aim of life is to become whole, which he said is to become conscious of the unconscious. You talk about how mind at large is trying to find this self-awareness or self-reflection. Yes, with the caveat that Jung used the word "consciousness" as a synonym of what we would call today "meta-consciousness," conscious metacognition. To be conscious entails having experience. To be meta-conscious, on the other hand, is to have experience and to know that one is having experience. This explicit introspective awareness is the outcome of what Jung called individuation, which I do see as the ultimate goal, not only of life, but of the universe itself, through us. [*Jude Currivan* agrees.*]

Could you use the word God to describe that evolution in conscious awareness? That is fine, but the word has a lot of historical tradition; it's a word that needs to be used very carefully because it carries a lot of baggage. If I just say, yes, it's God, then people might think there is a God that has a plan and knows exactly what it's doing at a metacognitive level. I don't think that is the case because metacognition is something that *we*, humans, have brought into existence in nature: *We* are nature's way of developing metacognition. God itself, as the underlying universal consciousness, is metacognitive perhaps only to the extent that *we* are aspects of it. I don't think there is a deliberate universal plan scheduled nicely the way you would plan a vacation. I think nature itself, beyond us, unfolds in a fairly instinctual manner according to the directions provided by its archetypes. That is why nature is so predictable and can be modeled in such a reliable manner by science. *People who have had NDEs like Eben Alexander* came away with feeling love and joy, but not logic.* I think that makes a lot of sense.

Let's talk about the evidence you develop in your books showing that idealism is accurate. Experiences like using psychedelics or suffering brain injury, or where the brain is impaired in some other way, seem to indicate there are experiences that are independent of neurons firing. I think living creatures are dissociated complexes of universal consciousness in a sense analogous to a person with Dissociative Identity Disorder with multiple alters, alternative personalities, and different centers of consciousness. *One alter could have diabetes and the other not. One alter could wear glasses and the other not. To me, that is the best evidence of the power of personality and consciousness over the physical.* There certainly is very objective evidence about there being dissociated alters of experience. We have objective fMRI and EEG scans of that.[2] I think this provides a very nice analogy about what may be going on at a universal level.

Universal consciousness dissociates itself into alters. The extrinsic appearance of an alter—that is, what an alter looks like from across its dissociative boundary—is what we call a living organism. You are an alter of universal consciousness, as is the tree, bacteria, etc. If ordinary brain activity reflects not only our inner experiential states, but also the dissociative mechanism itself—then, under some circumstances an impairment of ordinary brain activity should reflect impairment of that dissociative mechanism itself. Such an impairment correlates with an expansion of consciousness as experienced from a first-person perspective. If you are impairing the dissociative process, you are making the dissociative boundary more porous so things can flow in from beyond the boundary.

The filter is reduced. That's an analogy I use with care, but that's a valid analogy. If you're disrupting the dissociation, the boundary becomes more porous and you begin reintegrating things that you were dissociated from, such as a sense of self, memories and perspectives that you've not been able to access since birth, for instance. That would register as an expansion of awareness, which for materialism is very counterintuitive, if at all reconcilable. Under materialism brain activity is not merely the image of dissociated inner life, it is the *cause* of experience to begin with, so any reduction or impairment of brain activity should *always* correlate with a *reduction* of the richness or intensity of experience, not an expansion. However, what we see in a very wide variety of cases is this consistent correlation between certain types of brain activity impairment—be it through damage, bullet wounds, hypoxia, chemical impairment, psychedelics, or trance—and an enrichment of inner experience. That is at least suggestive of an idealist metaphysics and contradictory with a materialist metaphysics.

You point out that indigenous people had mystical experiences associated with hardships like finding water in the Australian outback. If life is the image of dissociation, the more you can secure comfort, the more you can enforce your ego, the more dissociated you will be. We've become very good at securing the boundaries of our dissociation. The entirety of Western civilization has evolved in this direction to make sure we have very comfortable lives. We live sheltered lives that secure very optimal ordinary brain function. Primitive cultures were exposed to more physiological stress, which I suspect could have compromised the dissociative mechanisms. They had more access to something beyond ordinary life because their dissociative boundaries were consistently more porous than ours.

They were also not as overwhelmed by culture as we are. We have enforced a global culture that tells us what to believe in. Authority figures in academia for centuries even have been telling us what is possible and what is not possible in principle. The culture of those preliterate societies didn't have a global civilization or global media reinforcing the same message again and again. I think they had access to aspects of reality that most of us in Western civilization don't even suspect.

Some people can access the bigger consciousness without something impairing the brain. We can meditate or go into trance, or I learned skills to do intuitive work. So, it's possible to reduce the filter in ways that aren't so hazardous. When I say impairment of ordinary brain function, I don't mean it in the sense of injury or toxicity. For instance, if you do a breathing exercise of hyperventilation, you will be impairing ordinary brain function because you will increase the alkalinity of your blood through hyperoxygenation. That increased alkalinity leads to constriction of blood vessels, which will reduce the amount of oxygen your brain gets and therefore will impair ordinary brain function. But that doesn't mean that it's toxic. *You can also intend, "I'm going to go into this other state and do telepathy or healing from a distance," just with intention.* I don't have experience with that. *I do it in clairvoyant sessions.*

Another proof of the limits of materialism is quantum mechanics, the observer effect, the entangled non-local particles, the connection they have even if they are separated at a distance, agree? Yes, but we have to be rigorous in how we interpret this because there is a lot of misinterpretation. The experiment you alluded to is the quantum entanglement experiment in which you have two particles that are sent apart over a long distance. The observation made on the first particle has an instantaneous effect on what is observed on the other particle, even if the particles are on opposite ends of the universe. There are two ways of interpreting this, both of which would contradict the vanilla, mainstream, materialist metaphysics. One way is to say that the universe is non-local with hidden properties of the universe "smeared out" across all space-time. They are not localized in the particles themselves because the particles would have to communicate at a speed greater than light and that is impossible. To coordinate what you get on either measurement, you would have to assume that there is a smeared out hidden property of the universe that, from behind the scenes, connects the two particles and determines what you observe when you make a measurement. That is non-locality. It assumes that the particles are observed in the same physical universe. This assumption is the motivation to postulate a non-local hidden property of the universe.

A second interpretation is to say the physical universe has no objective existence, called anti-realism. The idea is that things only become physically real upon observation, only when you make a measurement. Before you make a measurement, things don't exist in a physically real way. They exist only as possibilities in a realm of superposed possibilities that are not physical in the concrete way that we consider physicality to entail. They don't have a definite position in space-time or definite boundaries, properties, or mass. They exist only as potentialities, not as physically real entities, until you look at them and that potentiality collapses or translates into a physically real entity. For a number of technical reasons related to what we call Leggett's inequalities—which were confirmed in studies in 2007 and 2010—the explanation for this spooky action at a distance, for this correlation between measurements made at different ends of the universe, is that the physical universe is not physically real until we look at it.

What is out there, until we look, is something purely mental, in a superposition of thoughts or experiential states, similar to the way that we can hold contradictory thoughts at the same time in our minds. I think the inanimate universe at large exists only as a superposition of transpersonal thoughts until we interact with it. It's the act of this interaction between an alter and the inanimate universe that brings the physical universe into existence. If that is the case, my physical world is essentially different from yours because my physical world is brought into existence by *my* personal observation. We are immersed in the same ocean of superposed transpersonal thoughts, but it is not physical until we look at it. My physical world, although physically different from yours, is consistent with yours because it arises from my interaction with the same ocean of transpersonal thoughts that you are immersed in. If you say this, you no longer need non-locality—it loses all meaning. Carlo Rovelli proposed the relational interpretation of quantum mechanics in 1994. Although he is metaphysically agnostic, he is just saying what quantum mechanics implies. He acknowledges that there are some metaphysical qualms that arise from those implications but doesn't address those qualms. I think those qualms require an idealist interpretation of the essence of reality to be made sense of.

How can you say they are in separate universes because they are still connected? Bob will only know that his measurements are correlated with Alice's once he meets Alice, either directly or through some communication operating under the speed of light. Only then do Bob and Alice find out, retrospectively, that their measurements were highly correlated. In other words, only when the physical world of Bob "comes in contact"—that is, exchanges information—with the physical world of Alice do they realize that what they did was highly correlated.

They come in contact when they are measured or observed? No, if Bob is fundamentally separate from Alice on another side of the universe during the measurements, they have no idea that what they are measuring is correlated with what the other is doing. If they know it, it means that they are in communication at sub-luminal speeds and, therefore, the experiment is not valid. The experiment is only valid when there is no possible communication at the moment of the measurements. The measurements have to be made at the same time with a distance that is large enough to eliminate the possibility of communication between them. But when they come together and tell each other what they did, then they realize there was a correlation. But then their physical worlds are "in contact," that is, exchanging information. Alice always has her physical world and Bob always has his physical world. They are immersed in the same ocean of transpersonal thoughts which serves as an intermediary for messages to be sent across their respective physical worlds. It is a complex topic.

If the electrons weren't being measured, would they still change in response to the other's change of spin or is it the act of measurement which gives them their connection? Quantum physics means that if you measure spin on an electron and that electron has an entangled partner on the other side of the universe, then whoever measures the spin of the other electron will get a result that correlates with the first measurement. Fundamentally, what it says is that the underlying reality of those two entangled electrons cannot be looked upon as separate systems. The electrons are just different aspects of a single, unitary, indivisible system, even though they are in different physical worlds, so to speak. It is a statement about behavior that the entangled electrons behave in a way that cannot be separated. Their behavior is intrinsically unitary and indivisible.

To turn this observation of behavior into a conclusion about what the world essentially is, is an exercise in philosophy and in interpretation. My interpretation, which is consistent with the observed behavior, is that Alice and Bob live in two fundamentally different physical worlds. They do share an environment, but this environment is not physical. It is an environment of superposed thoughts, which becomes translated into a physical world upon their personal observation. The fact that Alice's and Bob's observations correlate means that there is a transpersonal environment of superposed thoughts that connects them. But what connects them is not the physical world of their measurements. It is the thing that gives rise to their respective physical worlds upon their mental interaction with it.

What connects them if they are in separate places other than they are being observed? When you talk about separate places, these are different places in one physical universe, but the physical universe of Alice is different from the physical universe of Bob. It makes no sense to talk about separate places for Alice and Bob because they are not in the same physical universe. What makes sense to say is that they operate according to separate cognitive perspectives within the ocean of transpersonal mentation in which they are immersed. But that is not a place, it's a cognitive perspective. Place entails space, which is a property of the physical world. Alice has a physical world and Bob has another. Within Alice's physical world it may be valid to infer that Bob is occupying another place. But in fact, he is not because he has his own physical world. What he does have in common with Alice is that he is immersed in the same transpersonal ocean of superposed mentation, of transpersonal thoughts. That difference in perspective becomes cognitively translated into the respective physical world of Bob as a different position in space. What connects them is transpersonal mentation, the mind of Brahman, the collective unconscious, the *akasha*, whatever name you want to call it.

The hard problem of consciousness is how conscious experience can emerge from unconscious matter since atomic particles are thought to lack consciousness. But what you've just explained is it's not really a problem because they are part of this universal mind. I think the hard problem is an invention of materialism. All we have is experience—our entire lives are made of a series of experiences. That is all we can directly know. Everything else is a theory and an inference *of consciousness.* It's consciousness trying to make sense of itself and wondering what is going on here. Why are these experiences unfolding with the patterns and regularities with which they do unfold?

Then consciousness postulates this abstract entity fundamentally beyond experience, which we call unconscious matter, a theoretical model that attempts to make sense of the regularities of experience. The problem is, after consciousness makes/creates this abstraction to explain things for itself, it tries to reduce itself to that abstraction. It then tries to say, "I, consciousness, am nothing but matter." It's like chasing your own tail at light speed or a painter who paints a self-portrait and then points at it and says, "I am that portrait!" Then this painter has to face the immediate implication, which is, "How does that portrait explain my conscious inner life?" Of course it doesn't since that portrait is a product of one's conscious inner life. That is the hard problem of consciousness: Trying to get the painting to explain the painter. That is why it is not hard—it is impossible. It's a logical artifact of a malformed and hopeless metaphysics that tries to reduce consciousness to its own abstractions.

Materialists are just chasing their own tail. The hard problem is a nonexistent problem because it arises out of an artificial and flawed line of thought to begin with. That is why it can never be solved. It's an invented problem. The legitimate challenge is to explain everything else in terms of consciousness, the starting point of knowledge. The challenge is to explain the patterns and regularities of nature in terms of the patterns of excitation of consciousness itself. That is what idealism tries to do.

You imply that consciousness doesn't stop after death? If consciousness is all there is, then life and death happen within consciousness. Consciousness has nowhere to go. It's what there is. It's the framework where things happen. Consciousness has nowhere to disappear into since it's the fabric of existence. *Personal* consciousness is another matter. If we are dissociated alters of universal consciousness, it's completely reasonable to imagine that this dissociation comes to an end. Death is what the end of dissociation looks like. What happens to the alters of a Dissociative Disorder patient when the patient is cured? They are all reintegrated into the host personality. The host personality may even remember the alters. Does the consciousness of the individual cease to exist upon the reintegration? No, it doesn't cease to exist because it's all there was from the very beginning and to the end. It's just a false sense of identity and the temporary interruption of cognitive associations that come to an end.

I would compare that to having a dream about a certain character and you think you *are* that character until you wake up and realize it was just a story you made up. Even the world in which that avatar existed was a product of your dreaming mind. I think death is analogous to that dream in that when the dissociation ends, you're reintegrated into the host universal consciousness. You may even remember everything that happened as a kind of dream. Now, I am back to what is truly real. *That is what Eastern religion says.*

In your book, The Idea of the World: A multi-disciplinary argument for the mental nature of reality, *you talk about hidden psychological motivations behind mainstream physicalism. Do they want to keep us unquestioning?* I wrote a psychology paper about it, published in the journal *SAGE Open*, which I think is the most-read academic psychology journal in the world. I had doubts whether I could publish this paper because it's meant to be a bit of a slap. Physicalists tend to psychologize anyone who is not a physicalist, as if they were crazy or wishful thinkers since only physicalism is based on facts and is objective. I thought that is complete nonsense because there are lots of ego-defense mechanisms woven into physicalism.

I think physicalism, in surprising ways, provides us with reassurance about death, since the unknown of the experiential state beyond life has been the greatest fear of humankind for millennia. "Am I going to go to heaven or hell?" That has been the greatest anxiety of human beings through most of history. Physicalism does away with that with one fell swoop when it says your problems are guaranteed to end. People say physicalism must be an objective perspective because it turns the universe into a meaningless mechanical contraption driven by pure chance and blind laws. It drains the meaning out of life so you have to be a very tough, objective person to acknowledge that the data shows us that physicalism is correct. Nonsense.

Physicalism provides great avenues for finding meaning according to a psychological process called "fluid compensation." Empirically established, fluid compensation is a psychological defense maneuver that we do when our meaning system is threatened. We derive meaning from a number of sources: from self-appreciation, from participating in something bigger than ourselves that will survive our death or from eliminating doubts. When you eliminate doubts you recover a sense of meaning. Suppose that things go terribly wrong in your life but you understand why it went wrong: You recover meaning from that.

Fluid compensation means if one of these sources of meaning is threatened, you will compensate by increasing your reliance on the other sources of meaning. You would do that "unconsciously," in a non-metacognitive way, almost instinctively. Religion is a source of transcendent meaning. If your religion is threatened, you will compensate by seeking meaning in another way. Mainstream physicalism is a product of 19th century thinking when, with the "death of God," scientists had to confront their own mortality. They didn't believe in God or the soul anymore, so their death anxiety increased, which is called "mortality salience." The intellectual elite, through fluid compensation, sought other sources of meaning, such as leaving major scientific achievements behind. Finding closure through eliminating doubts has been a major non-transcendent source of meaning for scientists. I worked at CERN in Switzerland, where we invested billions in a particle collider in order to close the standard model of particle physics. Currently, it has no practical applications and may never have. But we have closure, which gives us a source of meaning that replaces the meaning lost through rejecting God, soul and transcendence in general. That is fluid compensation at work.

The problem is that the average person on the street has no access to these alternative sources of meaning or understands the standard model of particle physics. That we found the Higgs boson does not increase their sense of meaning. That is the schism between people who have no fluid compensation, no access to other sources of meaning legitimized by materialism, but nonetheless have materialism imposed on them, versus the scientists who promote materialism. Their message destroys the meaning of the lives of non-scientists as reflected in the culture wars today. There are many other reasons why physicalism provides ego-defense, ego reassurance and why physicalism has a host of psychological wish-fulfilment motivations behind it.

In More Than Allegory *you say that a lot of religious myths are actually true and this inner storytelling helps us. What is an example of a true religious myth?* When I say "actually true" I don't mean that it's *literally* true. I think interpreting religious myth literally is killing it and is just as bad as dismissing it as untrue. When you interpret it literally it becomes just a story, an absurd and unrealistic story. I think that religious myth is true in that it's a *symbol* that points to an aspect of reality that is really true, but which cannot be corralled into direct, positive, analytic statements; it cannot be corralled into a literal model because it transcends the boundaries of logic. It transcends the boundaries of conceptualization insofar as our cognitive abilities allow us to develop.

What about the story of Adam and Eve who leave the garden of innocence and unconsciousness and then go out into the world as an allegory for how we have to experience to become self-aware. The beautiful thing about symbols is that they have multiple valid interpretations, or you wouldn't need them. You would just provide the interpretation that is true and you would be over with it. The beautiful thing about the symbol is that you have extra interpretations. A valid interpretation of the myth of the Fall—and it's the one I personally resonate with most—is that the Fall means the rise of self-reflection, the development of our ability to experience but to also know *that* we are experiencing. When Adam and Eve took a bite from the fruit of the tree of knowledge, they realized that they were naked. Presumably, they experienced their nakedness all along but when they took a bite from the fruit of the tree of knowledge they realized *that* they were naked, through metacognition.

The problem is that, once you become metacognizant, you begin to suffer, as you learn you are going to die. You begin to live in the past and all of its regrets, as well as in the future and all of its anxieties. That is the cost of metacognition, but it's very valuable because it allows us to develop a level of consciousness that would be impossible otherwise. Otherwise, we'd be immersed in the currents of instinct like a leaf in a tsunami, without being able to contemplate what is going on. Metacognitive consciousness gives us the ability to stick our heads out of the flow of water and think, "This is what's happening." That is very valuable but with it comes the suffering of regret about the past and anxiety about the future. We were expelled from the garden because we could no longer live in the blissfulness of the present moment, like the other animals in the garden, and we had to toil the ground. We had to suffer because we acquired the knowledge of the gods, the ability to metacognize. That is one way of interpreting it. I think there are many other valid ways. Even if you put them all together you still would not exhaust the symbol, because otherwise, it would not be a symbol.

In the face of climate change, the increase in autocrats, growing inequality around the world, knowing what you know about the universe, are you hopeful or pessimistic about our human survival? I am very optimistic about the survival of our species. Inuits and African Bushmen have the skills to survive and walk this earth. Even if we wreck this planet, some of us, like the Aboriginals in Australia, have the skills and have the connection with the planet to continue. I'm extremely optimistic about the planet. Even if we wreck it, give it a million years, which is a blink of an eye, and it's all good again. Even our species is going to survive. The question is: will our *culture*, our *civilization* survive or, as Noam Chomsky puts it, will "organized human activity" survive? I'm pessimistic about that because there is so little time left before the point of no return. I do not think we can turn this Titanic around in time to avoid the iceberg so the name of the game now is to plant enough seeds for those who survive to build upon.

What do you do for fun in the face of work, writing, girlfriend and cats? Girlfriend and cats are very important and a big part of my personal time is devoted to them and home life. I love my home. We have a vegetable garden where we produce up to 150 kilos of vegetables every year. My girlfriend is an artist, but her hobby is the garden. I tag along with her on that hobby, which is very enjoyable. I very much like contact with nature. Whenever I have the opportunity I like to go walk in the mountains or I go to the sea. I love a region in Germany, only three hours from where I live, called the Mosel Valley. I love Switzerland and the mountains and I love the North Sea coast. When I have the time, that is where you will find me, but I have very little time.

I work full-time as a corporate strategist. That is how I pay my bills and have this house. I read and write a lot. I have lots of other activities. I help out with a publishing house. I'm co-director of a charity organization that gives me personal fulfillment. I have a kazillion of activities that are not really hobbies but second, third, and fourth jobs. Contact with nature is what recharges me, as well as my girlfriend and cats.

Books

Rationalist Spirituality: An exploration of the meaning of life and existence informed by logic and science, 2011

Dreamed up Reality: Diving into mind to uncover the astonishing hidden tale of nature, 2011

Meaning in Absurdity: What bizarre phenomena can tell us about the nature of reality, 2012

Why Materialism Is Baloney: How true skeptics know there is no death and fathom answers to life, the universe, and everything, 2014

Brief Peeks Beyond: Critical essays on metaphysics, neuroscience, free will, skepticism and culture, 2015

More Than Allegory: On religious myth, truth and belief, 2016

The Idea of the World: A multi-disciplinary argument for the mental nature of reality, 2019

Decoding Schopenhauer's Metaphysics: The key to understanding how it solves the hard problem of consciousness and the paradoxes of quantum mechanics, 2020

Endnotes

1 http://grahampriest.net

2 https://www.tandfonline.com/doi/abs/10.3109/hrp.7.2.119?journalCode=ihrp20 DOI: 10.3109/hrp.7.2.119

Dave Pruett, Ph.D.

The Third Copernican Revolution

Photo by Mike Miriello

Questions to Ponder

How did Dr. Pruett resolve his personal conflict between religious and scientific beliefs?

What is the third Copernican revolution and how did it begin?

All of the "sacred cows" of the second revolution are dead except one. Explain.

What evidence convinced Dr. Pruett that we share a collective unconscious, noosphere or cosmic web?

I was born in Durham, North Carolina, where Dad, recently returned from World War II, completed his undergraduate education at Duke University and continued on under the GI Bill to earn a medical degree. Dad was the first of his family to attain a college degree, much less a medical one. Mom had only a year of college, having had to drop out because her widowed mother could not afford Mom's education. After Dad's internship at the Medical College of Virginia, we moved back to Bluefield, West Virginia, my parents' birthplace, so I grew up in southern West Virginia from the age of three until high school graduation at 18. Bluefield was very segregated, with a black part of town and a white one and separate, unequal schools.

Although I didn't grow up in Appalachia, you could see it from my doorstep. The topography of Appalachia—convoluted mountains with dead-end "hollers"—and a one-industry economy, coal—can severely restrict one's horizons. In contrast to the regional norm, I grew up relatively affluent, in a doctor's family. Going "home" feels like visiting another world, a world that seems closed in upon itself. One of the best portraits of life in the West Virginia coalfields I've encountered is Anthony Bourdain's segment on McDowell County, West Virginia, that captures life's daily deprivations in Appalachia, but also the resilience, grit, dignity, and tight-knit communities of those living there.

In terms of personality, I'm an INFJ on the Myers-Briggs personality inventory; that is, strongly introverted, with a profile characteristic of "authors." I am both highly rational and highly intuitive, which has led to considerable internal struggle, yet, that may also be my greatest gift. As to astrological sign, I'm a Cancer, but I've not paid much attention to astrology.

What was your family's attitude to religion and spirituality? How has yours evolved? My dad was not particularly religious but he had incredibly high ethical standards and a nearly infallible moral compass. As a physician, he was also rational and scientific. In contrast, Mom was a devout Southern Baptist, very intuitive, and a bright woman, but opportunities to grow never really came her way. As a result she held what I would call a narrow worldview circumscribed by the Baptist Church, the PTA (Parent Teacher Association), and the DAR (Daughters of the American Revolution). There was poor communication and considerable dysfunction in our family, partly due to Mom's precarious emotional state. Moreover, people who've had a doctor for a father can tell you that doctors are preoccupied and godlike. When I was younger, Dad seemed almost unapproachable for me, and Mom was so needy that I got sucked into her orbit, which was not healthy for my own emotional formation. Thus I grew up torn between two poles represented by Dad and Mom––the scientific/rational pole and the spiritual/intuitive one.

In college at Virginia Tech, I was active in the Baptist Student Union, and despite majoring in engineering and being in Air Force ROTC, considered going to seminary following graduation. Mercifully I couldn't get a deferment, for that would have been an unmitigated disaster. Until midlife I kept toggling back and forth between the competing poles, the scientific/rational and the spiritual/intuitive. Resolution was a long time in coming. As I have matured, I have come to distinguish spirituality from religion. For more than 30 years I have attended Quaker meetings because of the emphases on social justice, simplicity, and contemplation. I also find much to admire in Native American spirituality and Buddhism. I have learned to embrace the mystery without having to explain it all away. I think we are all individual waves in a vast sea of consciousness. At death, our separate wave ceases to exist and merges back into the sea. I suspect, based on accounts of mystics and NDEs, we will feel completely and totally loved at passing to the other side, whatever that entails. At least, that's my hope.

Why did you decide to study mechanical engineering? I grew up a child of the space race, enamored of science, particularly of space and rocketry, so my ninth-grade vocation paper focused on aerospace engineering. Mechanical engineering (ME) is similar but a tad broader, so I chose that as a major. However, I found engineering to be so narrow that I almost bailed out my senior year. Even though I loved math and science, I felt I needed balancing aspects to my education. A wise department head, not wanting to lose the student who was second in standing in the ME department, cut me some slack by allowing me to replace technical electives with humanities courses. That was enough breathing room to get me through the program.

After a hitch in the Air Force, I wound up teaching high school mathematics, having enjoyed tutoring while in the service. After three years of teaching, I went to work for an aerospace contractor at NASA Langley Research Center, the first of three times I worked there. I did indeed become an aerospace engineer, for a decade in all. In 1980 I earned a master's degree in applied mathematics at the University of Virginia then returned to NASA Langley. Two years later I enrolled at the University of Arizona for my Ph.D. in applied mathematics. Until that point in life, I had lived psychologically in the small world inherited from my mother, but in the completely foreign and wide-open natural environment of the Great Sonoran Desert, my worldview expanded by leaps and bounds. Some of that process of growth was intensely painful. Fortunately, I had access to those who could help.

After completing my Ph.D., I took an academic position at Virginia Commonwealth University (VCU). Although I loved VCU and Richmond, my first academic experience was not positive because of departmental dynamics, so in 1989, I returned to NASA Langley where I remained for nearly six years. I didn't arrive at James Madison University, my current affiliation and academic home, until the age of 47. I have been a late bloomer in every regard: Ph.D. at 38, married at 39, and father at 47.

While living in Richmond and teaching at VCU, I stumbled, with uncanny synchronicity, onto a mentor who deeply touched my life. His name was John Yungblut: Quaker, civil rights advocate, author, and lecturer. At the time, John was 71 and struggling with Parkinson's disease. After attending his lecture on Carl Jung, and later, an Advent retreat with John, I was hooked.

I asked him to serve as a mentor and he did so faithfully for the remaining ten years of his life. I dearly loved the man, a gem of a human being. John helped me pay attention to the competition between my spiritual and scientific poles and encouraged me to muster the courage to be faithful to both aspects rather than to choose one over the other, as I had been doing for years by so many flips and flops. He fostered this understanding by introducing me to towering examples of those who have successfully negotiated "the mystical struggle," including Carl Jung [*see his* Memories, Dreams, Reflections] and Teilhard de Chardin. Each of these intellectual giants had endured a tension of opposites that at times felt as if it would tear them apart. Yet resolving the tension gave rise to something transcending either pole. To paraphrase Jung's protégé, Marie-Louise von Franz, if one can live with the tensions of opposites long enough, one can give birth to something wholly transparent.

John introduced me to famous mystics and nurtured my own mystical impulses. A "mystic" is one who honors "knowing from within" (intuition) on equal terms with "knowing from without" (sensory perception and reason). Alternately, a mystic is one who is in awe of the universe. Thus, each of us is born a mystic. Unfortunately, too many lose the sense of wonder common among children. Early in my association with John, I awakened one morning in a state of excitement, pregnant with a poem that spilled onto paper, "Psalm of Resolution." It marked the beginning of a process of successful psychological integration.

Soon thereafter I began to fantasize about teaching an integrative honors course, having come to suspect that my struggle for integrity was reflected in the larger societal conflict between science and religion. The title of the course came to me in whole cloth: "From Black Elk to Black Holes—Shaping a Myth for a New Millennium." In 1999, that phantasy became reality at James Madison University, and at its second offering, the course received a science-religion course award from the Templeton Foundation. The overarching theme was to look at the universe from diametrically opposed perspectives—from a mythological perspective and a scientific perspective—and then to see if we could collectively bridge that gulf. More boldly, could we weave a new "myth of meaning" that is faithful both to the revelations of modern science and to ancient spiritual wisdom. The course, which I've now taught seven times, succeeded beyond my wildest expectations and touched the lives of many students, as it did mine. I consider it one of the most significant accomplishments in my life. From the course came my book *Reason and Wonder,* although it was 13 years between initiation and completion.

Was Yungblut a therapist? John was a spiritual counselor. He graduated from Harvard Divinity School and was an Episcopalian priest until the age of 44, when he had a recurring dream that his vestments were a straitjacket. John then left the ministry, became a Quaker, and moved to Atlanta to become involved in the civil rights movement. He knew Dr. Martin Luther King, Jr. personally. From that point forward, John never held what most of the world would consider bona fide employment. He lectured, wrote books, led retreats, and founded a non-profit called Touchstone, Inc., devoted to spiritual guidance. One of his books is titled *The Gentle Art of Spiritual Guidance.* His great gift was to uncover what an individual was passionate about and then to bless that passion. He was a radiant figure, and I always felt a brief afterglow from his presence.

How else have you coped with the difficult times in your life besides having a mentor? One of the most difficult times in my life was in January 1982. At 34, I experienced a terrifying major depression. In retrospect, it was one of the best gifts the universe has ever given me, because it marked a major step in a long-delayed process of growth. I'd tried far too long to live in the narrow world inherited from my mother. That tight garment no longer fit and it was crushing my spirit. The depression was a molting process by which I shed the old skin and grew a new one. I left Mom's orbit for my own orbit and established a much more satisfying relationship with Dad. Shortly thereafter I moved to Arizona for graduate school. There I met my wife, and a few years later, back in Virginia, I met John. My world opened up and became much richer.

From time to time, I've benefited from help from professional counseling and I've tried to gift myself with experiences, often conferences, that feed my spirit. For example, in 2000 I attended one of the first groundbreaking Science and Consciousness conferences in Albuquerque. There I met Huston Smith, and though I was only in his presence for about two minutes, it was a significant two minutes. In 2002 I attended an extraordinary workshop on dreaming in Sedona, Arizona. And I've gone to wonderful SAND (Science and Nonduality) conferences in recent years.

In your book, Reason and Wonder, *you talk about a third Copernican revolution. A premise is that consciousness has been left out of the materialist's paradigm of science. You give an example that if we perceive an electromagnetic wave of a certain wavelength, the brain translates it as red. Why red? Much brain activity doesn't simply record data; it shapes reality.* The overarching thesis of the book is that we can map the recent evolution of human self-perceptions through three successive Copernican revolutions. The first was due to Copernicus himself, who, in 1543, made a strong case for heliocentric rather than geocentric cosmology. This paradigm shift redefined the place of humans in the physical cosmos. Far from residing at the center of a cozy little universe consisting of a few planets and a few thousand stars, we exist in a universe of inconceivable immensity, comprised of 100 billion galaxies, each with 100 billion stars. Moreover, the universe is in continual flux. Today's Big Bang cosmology can be considered an aftershock of the Copernican Revolution, 400 years after the fact. Copernican revolutions thus contribute significantly to our origin stories and mythology.

Freud famously observed that Darwinism levied the "second blow to human narcissism," by which he meant that Darwin launched a second Copernican revolution. [On the Origin of Species by Means of Natural Selection *was published in 1859.*] Prior to Darwin, many humans believed that we had been placed on this planet through divine fiat, with humanity, created in the image of God, at the top of the animal kingdom. The science of evolution suggests different origins, a largely random process of successive mutations and adaptations over deep time. Moreover, we differ from the other creatures only in degree. We're now at the rough midpoint of the second Copernican revolution. When you poll Americans for the last 40 or 50 years, about half adhere to the belief that human beings were created by God within the last 10,000 years, a Biblical view of creation, despite extensive scientific evidence to the contrary. Copernican revolutions can take centuries to fully play out. The upshot of the second Copernican revolution—admittedly still underway—is to redefine the place of humans in the biological cosmos.

The first two Copernican revolutions were launched by individuals: Copernicus and Darwin. The third revolution is in its infancy and will likely redefine our place in the psychic/spiritual cosmos. It has to do with our understanding of the origins or emergence of consciousness (or spirit) in the universe and is a collective effort. Specifically, the relatively recent advent of quantum mechanics (QM) has turned our view of the material world on its head. The French philosopher Rene Descartes (1596-1650) divided the world into a physical part, the material world, and a thinking part, the conscious world. We call this the "Cartesian partition" of mind from matter. Descartes thereby paved the way for science to take the material world as its domain and for philosophy and religion to take the mental, spiritual, or psychic world as their domain. The Western world has lived with a sort of intellectual schizophrenia ever since.

Nobel laureate Ilya Prigogine asks: "Do we really have to make the tragic choice between an anti-scientific philosophy and an alienating science?" This warfare between head and heart, reason and intuition, science and religion is a genuine tragedy. Hopefully, the third Copernican revolution will help to heal this rift.

One of the most profound contributions of QM has been to collapse the Cartesian partition and the scientific belief in an objective observer who has no effect on the thing being observed. We now know that, at the quantum level, any attempt at measurement affects the state of the quantum object being measured. So mind and matter are not as disjointed as Descartes believed. This and other aspects of QM deeply bothered quantum physicists. For example, Einstein, whose explanation of the photoelectric effect, for which he received a Nobel Prize, contributed to QM, was vexed by the sort of random, non-deterministic aspects of QM. At heart, Einstein was a strict determinist. "God does not play dice!" he famously quipped. For example, if you have a lump of radium with billions of radioactive atoms, you can readily predict the decreasing radioactivity of the aggregate over time; it will decay exponentially. But the decay of a single atom is completely unpredictable, because it occurs randomly! Randomness introduces an element of uncertainty or indeterminacy into physics. Einstein believed in strict causality and determinism and yet QM introduced a non-deterministic element.

A similar thing happens when an excited atom decays and emits a photon of light, as studied by the science of spectroscopy. When you pass white light through a prism, you get the familiar rainbow of colors. However, if you heat an element, say, cesium, and pass the emitted light through a prism, you get a very different kind of rainbow, all dark except for brightly colored lines at specific places in the spectrum. Each element has a unique fingerprint in light, termed the Fraunhofer spectrum from its originator. One of the great successes of QM was to explain the Fraunhofer lines. In particular, Niels Bohr conceived of a new model of the atom whose electrons are constrained to occupy specific orbitals. When an electron jumps from one orbital to a lower one, it emits a photon of light of a particular wavelength or color that appears as a sharp line in the spectrum. When the electron jumps, however, is random.

Quantum jumping caused heartburn for many of the original quantum physicists. Erwin Schrödinger said: "If we are still going to have to put up with these damn quantum jumps, I am sorry I ever had anything to do with quantum theory." Similarly Einstein bemoaned (slightly paraphrased): If an electron is going to decide of its own free will, not only when it's going to jump, but where, I would rather be a cobbler or an employee in a gaming industry than a physicist. *So the fact of free will implies consciousness at play even at the subatomic level?* When Einstein uttered that statement above, he certainly didn't believe that an electron in an excited state had free will to choose when and where to jump. Over time, however, more and more reputable scientists have entertained the possibility that perhaps an electron does have a quantum of free will.

One of the thrills in writing my book was in 2009, when I got to visit Larry LeShan, a pioneering experimental psychologist, and Freeman Dyson, a pioneering mathematical physicist, then both nearly 90. At the time, Dyson was at the Institute for Advanced Studies (IAS) in Princeton, the intellectual "zoo" created to house Einstein when he immigrated to the US in 1933. IAS is still widely considered the world's premier intellectual think tank. In his autobiographical and highly regarded *Disturbing the Universe,* Dyson explores the interface between science and religion and entertains the idea of the actual existence of a quantum of free will. I asked him point-blank if he holds the view of quantum consciousness. He replied "yes," while freely admitting it's the minority view. In this view, however, there exists proto-consciousness throughout the cosmos, all the way down to the smallest quantum of the material world.

One finds resonant thought from the French paleontologist Teilhard de Chardin: "There is neither spirit nor matter in the world: the stuff of the universe is spirit-matter. No other substance than this could produce the human molecule." The conventional scientific view is that the universe is predominantly material, but that at a certain level of complexity, "poof," consciousness emerges. It makes far more sense to me that materialism and consciousness are both primary, that they are in Teilhard's view, two sides of the same coin. The so-called randomness in the universe may result from a degree of free will or consciousness at an elemental level.

I wonder if quarks have free will in some ways, because there are up quarks and down quarks, etc. What I'm suggesting about consciousness is not the common view among physicists. But at this early stage of the third Copernican revolution, it is difficult to predict whether or not the minority view might ultimately prevail. *Everything speeds up, as Alvin Toffler describes in his books.* Let's hope so. In truth, humanity is in deep trouble. We are in a tight race between environmental catastrophe and enlightenment. We really don't have centuries to become more fully enlightened. *It's a fact that electrons act randomly in atomic orbitals.* Yes, but we don't yet know whether randomness results from some kind of proto-consciousness involving free will or not. I don't think we're yet close to explaining consciousness with QM.

Along these lines, Larry Dossey* was a primary author of the recent "Manifesto for a Post-Materialist Science" signed by approximately 100 leading scientists. A lot of what we call "paranormal" is actually quite normal because it's ubiquitous. We call it "paranormal" simply because we don't understand it. The Manifesto advocates that science should open itself up to study these phenomena with the same toolbox used to study the material world. Dismissing such phenomena out-of-hand as bogus is anti-scientific because most of science's greatest discoveries have resulted from "damned facts," facts that refuse to fit the prevailing paradigm. But the Manifesto stops far short of implying that QM explains consciousness. To fully make or refute that connection could take two or three centuries of investigation.

What about quantum non-locality; the influence of entangled particles on each other even from a distance? In my book, I quote physicist Henry Stapp, who observed that Bell's Theorem, which has to do with non-locality, is "the most profound discovery of science." I met Stapp a few years ago at a Science and Nonduality conference. Quantum non-locality is experimentally based on Bell's Theorem, which in turn is based on an apparent paradox that Einstein and collaborators conjured up in 1935. It's called the EPR paradox, where EPR stands for coauthors Einstein, Boris Podolsky, and Nathan Rosen. In physics, mass is conserved, as are energy and momentum, and the paradox involves the conservation of momentum. Einstein and collaborators realized that certain quantum processes can give rise to two photons that travel away from one another, each at the speed of light. According to the current interpretation of QM, each photon has spin, a type of angular momentum. For photons, spin and polarization are the same thing.

A mind-boggling attribute of QM is the notion that the physical states of quantum objects exist only as probabilities until a measurement is made. Thus, each photon's spin (up or down) exists only as a probability, not as a certainty. If one then measures the spin of one photon, in order for total momentum to be conserved, the twin photon has to adjust its spin as well. Once the state is known for one, it is determined for the other. Einstein felt he had found QM's Achilles' heel, for it would apparently require superluminal (faster than light) communication between these two photons for one to adjust its spin in response to the measurement of its twin's spin. But such faster-than-light communication was forbidden by Einstein's own relativity theory.

For three decades, EPR remained a full-blown paradox. The resolution waited until 1965, when John Stewart Bell, an Irish physicist, proved a mathematical theorem about so-called "entangled" particles. The theorem stated that if Einstein were right, one set of correlations would hold between the spins of twin particles or photons, and if QM were right then another would hold. Measuring the correlations could tell you who was right. Around 1971, researchers at Lawrence Livermore Laboratory realized that an experiment based upon Bell's Theorem could actually be performed and the correlations always bear out QM.

There are several ways to ponder these astounding results. One is that there is superluminal communication between separated particles. Another interpretation is that space is a mental construct so that the twin photons aren't really separate. They just appear to be. I'm struck here by the wisdom of the ancients, such as in the three Sanskrit words (*Tat Tvam Asi*) meaning "that art thou," or "thou art that." This revelation is the core tenet of all religions: there is no separation between self and other. Now, in modern science, the separation between the object and the observer is also an illusion because you can't complete the experiment without the observer as part of the experimental apparatus. Moreover, non-locality suggests that one has to consider entangled particles or photons as a single system in which the apparent distance between the components is artificial in some way. *One of the scientists I interviewed said a possibility is precognition, in that one entangled particle precogs very quickly that the other is going to change its spin.* Yes, that's another, equality intriguing possibility. But we should be cautious to not prematurely conclude that one interpretation prevails, such as that QM implies precognition.

What about the evidence that there is something other than material?

You quote Schrödinger who says thoughts and feelings are outside of matter. Mind is an intangible ghost. We need a new paradigm for how to relate the brain to the mind. You mentioned telepathy, precognition, clairvoyance, distant healings, OBEs and NDEs. How do we know that those phenomena really exist? First of all, such so-called paranormal phenomena are ubiquitous. *The phone is ringing and I know it's my mother, right?* Exactly! Or a mystical experience, or an OBE, or an NDE, such as that described by neuroscientist Eben Alexander,* or precognitive or telepathic dreams. I, for one, have had a few of those dreams in my lifetime.

Let me give an example of what happens when one opens up to the world of the paranormal. I mentioned meeting Larry LeShan, a pioneering experimental psychologist. Two of his books especially spoke to me: The first was *The Medium, the Mystic, and the Physicist: Toward a General Theory of the Paranormal.* It was written in 1966 and so prescient that when re-released in 2003, the only thing Larry changed was the preface. The other book, a little classic titled *How to Meditate,* has sold more than a million copies. Of the latter, Larry confided that the book wrote itself; it just flowed out of him.

When LeShan was a newly minted Ph.D., feeling his oats, he intended to shoot down all this paranormal "nonsense." But in his own words, "he made a mistake." He looked at the evidence, mountains of it collected over a century. Because people are so skeptical of paranormal phenomena, most experiments in parapsychology are far more carefully controlled than is commonplace for routine psychological studies. He shifted from wanting to debunk to wanting to understand, and in the process found his life's work. Larry first chose to study paranormal healing and he's been credited by some as the father of mind-body medicine. Over the years, he and coworkers developed a meditation-for-healing workshop, which I attended in 2005.

To Larry, it's absolutely clear that there are some people who have healing gifts. One such person was a psychic healer by the name of Eileen Garrett. Larry studied her technique carefully. She was the "medium" in the title of his book. When Garrett was in a healing modality, she entered into an altered state of consciousness, a state reminiscent of mystical awareness described in literature over the ages. Larry also made friends with Henry Margenau, a well-respected physicist at Yale University. Once a week, LeShan and Margenau met for freewheeling discussions, including those regarding parallels between mysticism and modern physics. They accumulated a considerable number of quotations, some from modern physicists, others from modern or ancient mystics. They scrambled the quotations and gave them to Margenau's graduate students to identify who said what—physicist or mystic? Students got it right only about half the time, indicating there is considerable overlap between how ancient mystics and modern physicists describe the inner workings of the universe, with the common recognition that the world we see is largely illusion.

You also mentioned his book A New Science of the Paranormal. Yes, LeShan wrote this when he was about 90. He confessed to me to feeling like a jilted lover because the love of his life, psychology, had been philandering. In the 1970s, psychology was hijacked by the materialistic paradigm and largely taken over by B.F. Skinner's radical behavioralism. Skinner believed that free will was an illusion. Larry's book is not so much a comprehensive theory as a cry for psychology to take a deep look at itself and to find its better angels. *Especially now when psychiatrists rely so much on pharmaceuticals, just dealing with the symptoms.* Yes, psychology has gone downhill a lot since William James.

The most convincing demonstration of a collective unconscious of which I am aware is the Dream Helper Ceremony (DHC) originated by psychologists Henry Reed* and the late Robert Van de Castle. I participated in this unique process six times. The DHC is a protocol for intentional dreaming for altruistic purposes, normally conducted in groups of eight to ten people. Multiple groups can be conducted in parallel, i.e., simultaneously. A typical format would be to set up the DHC on a Friday evening and then to meet the next morning to process what the dreams reveal. An individual is chosen as a "focus person" who is currently dealing with some difficult life situation: relational, vocational, medical, etc. The focus person does not reveal the specific issue at this time. Members of the group each promise to have a dream overnight for the benefit of the focus person.

This sounds absurd, but experience bears out two astounding facts. First, if you randomly ask people on the street how many remember a dream from the previous night, it's usually around 20 to 25%. But once you have promised the focus person that you're going to dream for them, the incidence of remembered dreams rises to in excess of 95%! That's phenomenal in its own regard. The first time I participated in the DHC, I had two of the most profound dreams of my life, both prescient and precognitive dreams. Second, no one ever asks the obvious question: How do I dream for another? At some level then, we intuitively believe that we share some subconscious connection.

The morning after dreaming, groups reconvene to mine the dreams for collective wisdom. More often than not, a mosaic emerges from shared dreams that goes to the heart of the focus person's issue. After all dreams are shared and discussed, the focus person reveals the issue and the dreams are mined again for wisdom. The therapeutic value to the focus person is extraordinary in most cases because they have felt supported without judgment or reservation by a group of altruistic strangers.

We see there are many suggestions from different quarters—QM, paranormal experience, the DHC—that consciousness is pervasive in the universe and that there is a collective aspect to consciousness. Many great physicists have recognized this likelihood. For example, Wolfgang Pauli has argued for an active role for consciousness in physics. And Eugene Wigner said: "Consciousness must be introduced into the laws of physics."

In your book you mention Dean Radin's 1997 book* The Conscious Universe. *Who are other key thinkers in creating the third Copernican revolution?* Yes, Radin is a leading light in the field of paranormal psychology. Other key thinkers include Freeman Dyson, who believed that elementary particles do have a quantum of consciousness. Sir Arthur Eddington, who helped to validate Einstein's General Theory of Relativity, said famously, "The stuff of the universe is mind stuff." Eddington also wrote an article titled "Defense of Mysticism." The mathematical physicist Sir James Jeans is famous for the quotation: "The universe begins to look more like a great thought than a great machine." These people are not flakes; they are among the deepest of thinkers. Other giants who suggest that mind is not simply an epiphenomenon of matter are Erwin Schrödinger, who wrote the little classic *Mind and Matter* and physicist Roger Penrose, friend and colleague of Stephen Hawking.

Regarding consciousness itself, the focus of the third Copernican revolution, the old, materialistic paradigm may be crumbling. Science has been asking this question, largely without success: How does consciousness, which is immaterial, arise from a material universe? It may be the wrong question. The German philosopher Immanuel Kant believed that he had precipitated a Copernican revolution by turning that paradigm on its head. He asked instead, "How does the representation make the reality possible?" That is, how does consciousness give rise to the material world? So we have competing paradigms: matter as primary vs. consciousness as primary. A third paradigm appeals to me, whose chief proponent was the French paleontologist Teilhard de Chardin. Teilhard believed that matter and consciousness (or spirit) coexist as flip sides of the same coin, which he termed the Without and the Within. This consciousness/spirit extends all the way down to the smallest quantum of spirit-matter.

So the chicken and the egg came at the same time? Yes, exactly! That's Teilhard's point of view. Science studies the external face, the Without. But there's also a Within that science, thus far, has largely failed to recognize. I resonate with Teilhard's thinking because it is congruent with Native American spirituality, which grants spirit to everything: humans, animals, plants, even rocks. Such a view is often belittled as primitive "animism," but to me it fosters a worldview that treats all of creation with the utmost respect. Everything is Thou. Nothing is an It.

Teilhard's view of the cosmos is quite exhilarating and deserves attention. He was both a paleontologist of the first rank and a Jesuit so devout in his faith that he prayed to die on an Easter Sunday, a prayer answered by a massive heart attack on April 10, 1955. As a paleontologist, he fully embraced evolutionary theory, recognizing that the entire cosmos is evolving in a process he termed *cosmogenesis*, signifying a universe in continual creation. When was the moment of creation? It's now! Cosmogenesis is Evolution with a capital "E" and is to be celebrated.

If you look at the universe through the wide-angle lens of Teilhard's cosmogenesis, you see two prevailing megatrends. The first trend is that the physical cosmos is running out of steam. The stars will eventually burn out, and the universe will end with a whimper in what thermodynamicists call "heat death." Specifically, the second law of thermodynamics requires entropy to increase over time, meaning that the energy in the universe will become gradually less available and uniformity will ultimately prevail. Dying stars will collapse, some into black holes, where energy remains imprisoned and inaccessible. In a word, the physical cosmos is decaying.

But if you look at the biological cosmos, exactly the opposite is happening! For this process, Teilhard coined the term "complexification" or "complexity-consciousness." In the biological universe, we observe that over time lifeforms are becoming ever more complex and that complexity is associated with higher consciousness. For example, the Big Bang starts with elementary particles that merge to form atoms. Later, atoms collect to form molecules and molecules form cells. The cells give rise to RNA, which gives rise to DNA, which enables both replication and mutation, which gives rise to biological evolution with ever more complex organisms.

Teilhard would part company with the evolutionary biologist Stephen Jay Gould, who famously wrote: "Evolution is purposeless, nonprogressive, and materialistic." On the contrary, Teilhard saw a direction imprinted upon biological evolution toward greater complexity. And as Sir Charles Sherrington noted: "With life comes mind." With greater complexity comes higher consciousness. Given that greater awareness affords an adaptive advantage, the direction of evolution has an inherent bias. Although evolution is largely governed by random events such as mutations, there is an arrow to evolution that points in the direction of complexity-consciousness. The two trends within cosmogenesis appear to function counter to one another in that the physical world is running down while the biological world is running up. The $64,000 question: Is there some relationship between the two megatrends?

The answer is yes! The latter could not happen without the former. We've only been able to make that connection scientifically since about 1963, when the new mathematical theory of dynamical systems, commonly known as chaos theory, appeared on the scene. It's complicated, but under certain circumstances, a decaying universe can produce small pockets, like eddies in a stream where water flows upstream, that defy the general trend. Within these pockets or eddies are the right conditions for self-organization and complexity to emerge. Specifically, Ilya Prigogine, a Nobel laureate in chemistry, studied self-organization in chemical systems. His findings are illuminated in a book for the lay reader titled *Order Out of Chaos.* Prigogine and coworkers identified three conditions that can lead to emergent self-organization. To be precise mathematically, the conditions are: 1) the governing dynamics are mathematically nonlinear; 2) the system is dissipative; and 3) the system is far from equilibrium.

Life satisfies all three. Biology is based on chemistry, which is strongly nonlinear because of catalytic reactions and feedback loops. All living beings function metabolically as "heat engines" and produce waste heat and are hence "dissipative" systems. Finally, the entropic eddies have to exist "far from thermodynamic equilibrium." Using an analogy of a stream, eddies form in rapid flows such as in the mountains, not in placid, flatwater flows near the coast. Earth, bathed in the radiant glow of the sun, is thus in a far-from-equilibrium state; that is, in the strong current of entropy. Thus, a generally decaying universe can indeed create pockets in which entropy can locally decrease; that is, pockets in which self-organization and greater complexity prevail.

You write that evolution is increasing biological complexity and this fits into the whole philosophical change we're undergoing. A chapter in a wonderful book by Peter Atkins titled *Galileo's Finger* clearly illuminates these processes. Atkins concludes that we couldn't have the upward running of the evolutionary stream without the downward collapse of the physical cosmos because the running down of the physical world actually propels the running up of the biological and conscious worlds. It offers an almost teleological view of the cosmos as a factory for higher consciousness.

Teilhard expresses this idea in poetic terms. He depicts cosmogenesis as a distillery of consciousness. He compares complexity-consciousness to the process of bees making honey from the nectar scattered about in flowers. Similarly, Complexification gathers scattered proto-consciousness into beings of higher consciousness like bees collect scattered pollen to make honey. Both are counter-entropic processes driven by the deterioration of the physical cosmos.

Teilhard also coined the term *noosphere,* as a sequel to the notions of geosphere and biosphere. Think of these concepts as concentric shells, like stacked Russian dolls. The geosphere is the physical earth. Surrounding the geosphere is the thin shell of the biosphere containing all life. Surrounding the biosphere, in Teilhard's view, is the noosphere, the sphere of collective consciousness. Teilhard died in 1955, long before the Internet, but in many ways it seems the Internet is the physical embodiment of his notion of the noosphere.

Chaos theory, which plays a major role in giving a scientific basis to Teilhard's complexity-consciousness, also spells the death knell for one of science's sacred cows: determinism. Since the scientific revolution burst onto the scene with the publication of Newton's *Principia Mathematica* in 1687, mainstream science has, over time, adopted a number of assumptions about how the universe operates, call them science's "sacred cows." These include materialism, locality, absolute time and space, dualism, realism, determinism, and reductionism. It is interesting to note that *all* of the sacred cows have died, save for materialism, and it's on its last legs. Yet science is alive and well, which implies that the sacred-cow assumptions are unnecessary.

Determinism is a sacred cow to which Einstein clung tightly. Chaos theory and QM have combined forces to kill off determinism. Determinism is the belief, rooted in a materialistic view, that the future is completely predictable from the laws of physics. In this worldview, the universe exists as a collection of particles whose motions can be predicted on the basis of laws of motion and knowledge of the initial positions and velocities of the particles. However, the Uncertainty Principle of QM states that one can't know the initial conditions of a quantum object with absolute precision because anything that one does to determine the velocity will contaminate the determination of position, or vice versa. Moreover, chaos theory introduces the concept of "sensitive dependence to initial conditions," known generally as the "butterfly principle." Under certain conditions, if you change initial conditions the slightest bit, you change outcomes enormously. So, if you can't precisely pin down those initial conditions, then you cannot predict what the future will be, because the future depends sensitively on the initial conditions. Modern science and mathematics have rung the death-knell for determinism.

How does consciousness fit in here? If events in the universe are not strictly deterministic, they must be contingent. If the inherent randomness in the universe is actually a manifestation of proto-consciousness, then consciousness is fundamental. A demonstration of this is Dean Radin's* and Roger Nelson's* experiments with RNGs. Both experimenters have demonstrated the small but persistent effect of conscious intent on the output of RNGs, particularly when many experiments are combined into meta-analysis. If consciousness can affect something at the quantum level, but the indeterminacy of the quantum level determines what's going to happen in the long-term evolution of the universe, then consciousness must somehow be fundamental to how the universe evolves.

Another dead sacred cow is Newton's hypothesis that time and space are absolute, the same for all observers. Surprisingly Einstein's special relativity found that time and space are not independent but intertwined in some intricate way. Moreover, from Einstein's general theory, we learn that the geometry of space-time is not Euclidean like a sheet of graph paper but warped by the presence of massive objects like stars. But the warping of space-time is also dynamic. In the great cosmic dance, matter tells space-time how to warp and space-time tells matter how to move. In this dance, the motions of stars or the interactions of black holes send gravitational waves rippling throughout the space-time web. So space-time is more like an interactive web than a stage. The events happening in space-time influence space-time and vice versa.

For an entire century, Einstein's predictions of gravitational waves, ripples in the space-time web, remained purely hypothetical. Then in 2016, gravity waves were actually detected by the LIGO experiment, vindicating Einstein![1] Once again, I find parallels between Native American mythology, which speaks of the "web of life" and the "cosmic web," and modern science. We live in a cosmic web. *The web is not independent of us and we're not independent of the web. It's all subtly interconnected.* [*See Rollin McCraty.**]

You say that love is woven into the fabric of the world. Yes, recall that, under the mantle of cosmogenesis, we observe two megatrends running counter to one another: The decay of the physical cosmos and the running up of the biological cosmos. The latter could not happen without the former, so it's almost as if the physical world is sacrificing itself in some way for the biological and conscious world. In human terms, we have a name for such counter-entropic, sacrificial activities, like home-maintenance, gardening, restoring an old car, or parenting. Each reverses the normal tendency toward decay. We call these activities labors of love, or simply "love" for short. It would almost seem that the decay of the physical cosmos could be construed as a labor of love for the sake of the conscious cosmos. Admittedly, this is an anthropomorphic view, but I find it a satisfying cornerstone for my own personal mythology.

As a productive person, how do you find time for fun? I'm not a high-energy person, nor particularly creative, so my productivity has come from a long, hard slog rather than flashes of brilliance. That said, the natural world has been my salvation, since my early teens. I love hiking, camping, biking, and walking. Since retirement seven years ago, I have been section-hiking the Appalachian Trail (AT) in Virginia and in August, I completed its 550 miles, one-quarter of the whole AT. I also play guitar and sing vocals in a wonderful band of older friends—what salvations are music and nature. I am also a contemplative by nature who enjoys meditation but I am not nearly disciplined enough to have a regular meditation practice.

Do you have another book that you're thinking about? Having just finished hiking Virginia's AT, I was inspired to work on a photographic essay of the experience and I may have landed a publisher. Once upon a time, I wrote some passable poetry and am toying with the idea of publishing a volume of poetry.

In the face of climate change and the rise of autocrats around the world are you optimistic or pessimistic? I am reminded of the quip by an early quantum physicist who said that on Mondays, Wednesdays and Fridays we teach the particle theory of light, and on Tuesdays, Thursdays and Saturdays we teach the wave theory. So, my answer depends on which day you ask the question. There's so much to be optimistic about and there's a lot to be terrified about. So many good people are awakening and working their hearts out to make a better world. And so many others, whether by ignorance or design, are doing everything in their power to keep us locked in an unsustainable status quo. In my optimistic moments, I hope that all the reactionary forces are simply the last, dying gasps of the old paradigm. Then, maybe *Homo sapiens* is just a stepping stone to something better, because truthfully we're screwing up royally right now. Where I'm most pessimistic is as a father. I have a 24-year-old daughter who is everything a father could possibly wish for. She deserves a bright and viable future but there are so many clouds on her horizon. Ultimately though the only viable option is to consciously choose hope and to work one's heart out.

Your additional insights for readers? Pay close attention to what your heart is telling you. As the French mathematician Blaise Pascal observed: "The heart has its reasons, which the mind knows not thereof." And this additional wisdom from my mentor John Yungblut, paraphrasing Jung: "We ignore our gifts at the peril of our immortal soul."

Book

Reason and Wonder: A Copernican Revolution in Science and Spirit, 2012

Endnote

1 https://www.nytimes.com/2016/02/12/science/ligo-gravitational-waves-black-holes-einstein.html

Harald Walach, Ph.D.

Doing Parapsychology Research in a Connected Universe

Photo by Anja Jahn

Questions to Ponder

What is Dr. Walach's Generalized Entanglement Model?

What is the most problematic aspect of psi research despite extensive evidence?

More qualitative and first-person research is needed, as is a circular model for research. Explain.

How did the non-materialist scientific paradigm develop in recent history?

I was born on the 6th of February in Augsburg, Germany. *Many of the people that I've interviewed are Aquarians, which fits because Aquarians are future-oriented and visionary.* I'm not particularly interested in astrology and I don't know much about it. *Do you know your Myers-Briggs personality type?* No, I'm not interested. *But you're a psychologist! My hypothesis is that these leaders in creating the new paradigm are intuitive rather than sensing and so far that's been true.* That sounds true for me as well, as I'm more intuitive than rational. I don't believe much in those typologies. I did my psychotherapeutic training so I know a lot but I don't find those typologies very helpful. I do know myself well enough to know that I'm kind of intuitive.

Do you find that when you're working with students in different countries that you approach them differently? I've lived in Germany, Switzerland, and England and now I'm back in Germany and teach in Poland. I see different mindsets or different approaches to education in England and in the German-speaking countries. In Germany, education is more broadly conceived, at least at the time when I was younger and teaching at university, whereas in England the young people are quickly funneled into subjects they prefer. A student could do A levels in England with no subject matter in languages or math, which wouldn't be possible in Germany, Switzerland and Austria. *In Germany you go to gymnasium if you're headed toward university while you go to a technical school if you're not.* That's right. In our A level, which we call "maturity," you have to include at least one foreign language, German, and a natural science. You can't bypass the major topics whereas in England you can.

You have two Ph.D.s in clinical psychology and in theory and history of science. What about your upbringing led you to be an academic achiever? I was actually more of a lazy guy but I happened to not find school very demanding so I liked it. I did a lot of music, I had a lot of social engagements, so school was more what I did on the side. *Why did you decide on psychology as your field?* I was always interested in people. Initially I wanted to be more like a healing person or a therapist.

How were you influenced by the religious context of your upbringing? My family was nominally religious as you were in Bavaria, a

Catholic region. I was educated at a school led by Benedictine monks in Augsburg. When I was 15 or so I dropped all forms of religious creeds and belief systems and was an atheist for several years. Then I discovered a meditation group next to our school, where I had my own experiences and that brought me back to spirituality and to understanding religious teachings from a more inner perspective. I dropped formal religion and rediscovered it for myself. *What form of meditation did they teach?* It was taught by a group of progressive nuns who included yoga, a form of Zen meditation, and imaginative meditations, including listening to music. I settled eventually on a Zen kind of meditation. I also did some meditation along the tradition of St. Ignatius [*16th century Spanish mystic and founder of the Jesuit order*] for about five years until I found that was too much of an imaginative type of work for me. I settled with a very generic Zen type of meditation.

I think of Zen as sitting and trying to keep your mind free of thoughts or working on a koan, *an unsolvable problem to keep your mind from going off in a logical direction. How would you describe Zen meditation?* It's a very generic concentrative meditation using a syllable to concentrate on, like a mantra called a *wato*. It's not just the meditation, it's also important to let the fruits of meditation grow in the garden of ordinary life. *Doing meditation and focusing on spiritual aspects of life do you find that you get ah-ha experiences downloading information? I think you refer to it as popping up out of the contemplative practice.* I would say it's a generic source of inspiration that comes along often, mostly when I'm not thinking of it, because meditation practice is like training your mind so that when you need it you are ready. Those inspirations they come along as I need them; sometimes they come at very unexpected times or during dreams. I wait until the impulse goes away, or if it doesn't go away, I take it seriously and follow it.

When have been the difficult times in your life and how did you cope? I've had a lot of difficult times in my life and the way to cope is to have my inner compass that I developed from my meditation practice and not to get too disturbed about it. It does upset you when people stop your funding, for instance, and you get the impression that there is no money coming along your road after the next two months. I've had that situation various times. My experience was by just trusting that I was on the right path, that something would come along and so it did.

Some people say having children is difficult for the relationship between the parents and having a baby wake you up through the night, as well as difficult dealing with teenagers. Was that part of your experience? No, it wasn't true for me although we have four children. I think it makes your life richer, more real, and more earthbound because you have to deal with ordinary things, wipe away the shit and make a child that is suffering from pain calm again. Of course it distracts you from what you might want to do otherwise. But it also helps you to appreciate the time that you have and funnel you into concentrating on what needs to be done. *You had a busy household!* Yes we did and I liked it.

What's an example of an insight in terms of your research where you came up with information that surprised you? Maybe the most generic thing, which I'm still working on, is the idea to generalize the entanglement relationship, which we know from quantum mechanics, to systems and situations outside of physics. That was something that popped up in my mind quite early. I was doing a lot of reading when I had these paradoxical data sitting in front of me, trying to make sense of them. Then it popped up in my mind that we might have a very similar relationship in other systems like those I was studying in complementary medicine, in homeopathy, or spiritual healing. That came out of the blue and I thought about the ramifications and spoke with people about it until it materialized in a more formal sense. Normally what happens if such an idea is really fruitful, I see a lot of synchronicities happen. I meet people who help me onward and have discussions with friends that point me toward a certain direction--or discussions with people who are not friendly.

This happened when I met the Chair of Theoretical Physics in Freiburg, Professor Hartmann Römer. He was present when I gave a talk on these issues in a philosophy seminar. He said, "Good idea, why don't we follow up on that?" We wrote a paper that prompted him to write about the formalism of generalized quantum theory. This brought in a couple of other authors and people interested in the idea. Normally when the ideas that pop up have some value, they meet up with occurrences in the outside world that make them grounded and realizable. If that doesn't happen then I give them up or let them sit on the shelf until I find an opportunity to make them real.

It's tempting to say with non-local entanglement we could conclude there's an information field, which would explain why we can do telepathy and ESP and prayer from a distance. But physicists would say no, we don't know what causes that kind of connection. I think physicists are correct to be really careful there. We are not calling this a physical entanglement because in quantum systems it is a very specific kind of state in which the system needs high levels of preparation, isolation, and a very artificial system that is very difficult to prepare. When I'm using the word "entanglement," I'm talking about a generalized form which may only be a metaphor for what we're experiencing, using the knowledge of physics about those entangled states to try to understand them.

I think the term "mechanism" is wrong because it's a word that comes from a Newtonian physicalist framework. It's more like a correlated occurrence of things that seem to be meaningfully related and somehow connected but we don't know how. So the term "synchronicity" is helpful because Jung was thinking of correlations between mental and physical occurrences without any further cause of connections between them. I think that is what is actually happening but we don't have a good theoretical model for that. Our model of generalized entanglement is one way of looking at things--more like a metaphorical way, certainly not meant to use the physical entanglement between elementary particles as an explanation but more as an image for what is happening.

What process allows ESP, clairvoyance, healing at a distance--those kind of phenomena where minds connect through space--without an obvious way of connecting them? We don't know. The model which I have tried to sketch is a generalized way of entanglement, which would mean a correlation between systems that is built up as they are set up. A decisive element is probably consciousness, how we partition the world, what we engage with that makes up a system. Sometimes we also use ritualistic demarcations of those systems. For instance, if you create a prayer ritual you do something to connect with a person by imagining her or him or you mention their name. That is the kind of ritualistic preparation of a unitary system of yourself, your consciousness, and the person you're praying for. One way of looking at it is that our consciousness is getting engaged there and may be able to produce a state of mind that somehow, by the way the system is set up, translates into something correlated in the other person's field of consciousness.

So this act of connection in prayer requires intention? Yes, it requires a certain mindset. I don't think this happens by accident. The way I'm conceptualizing it is a more parsimonious one because I don't want to postulate any particular fields, particles, or energies. I'm saying that maybe the way we partition the world and set our systems enables us to become part of that system, which enables us to either extract or experience information that is part of that larger system or be influential within that system by the way we form our consciousness. Other people might say that in order to really scientifically understand, you need to build up a hyper-spatial model with X dimensions in addition to the four we're living in. The model I prefer as a more parsimonious one is not necessarily the correct one.

We need to define consciousness if that's the core of everything. Consciousness is really a complex issue. One way of looking at it is its relational aspect as we always are in relation to something. For instance, being conscious of what we perceive and what we intend to do. Normally we define consciousness as some stance that is related to something else. That is usually the case but you can also have pure consciousness as in a meditative state where you're not relating to anything else but just to reality in a non-dual state. Sometimes consciousness can also be contentless, just pure presence, an object as it were.

You got into controversy for working with a master's student to research how telepathy works. What was that about? I used to run a study course for medical doctors, a postgraduate master's degree course for doctors training them in cultural sciences and complementary medicine. I was running a class in research methods and one of the students was interested in the idea of Kozyrev's Mirror. Kozyrev was a Russian physicist. He developed an alternative world model in which time is a variable we can manipulate. For instance, in quantum physics time doesn't even enter the equations. In relativity theory, time is one of the four dimensions and it is fixed by the limit of the speed of light. Kozyrev developed an alternative way of looking at the world and in his model time is a variable that we can manipulate. People who are interested in Kozyrev's theory use what is called a Kozyrev Mirror in order to manipulate time. We discussed how could you test that theory and developed a little experiment to create a miniature Kozyrev Mirror. It was fitted into a little box and connected with cables.

His hypothesis was that people who would be connected to that mirror would be better able to telepathically guess what content was in there. He put paper slips with numbers from 0 to 9 hidden in envelopes in the closed box connected to cables. Then he made a triple-blind telepathy experiment that was quite good. He ran about 600 experiments and found that there was only one situation where people were guessing better and that was when they could see that the cables were attached to the cylinder without knowing which numbers were in the envelope. It was still blinded but they saw they were attached. So in that situation the probability of guessing the right figures was more than you would expect by chance. It wasn't an incidental finding. The problem was the poor write-up of the unpublished master thesis. The skeptical community was very critical about the report because Viadrina University in Frankfurt was the only public university in Germany offering a complementary medicine course like that, which created a lot of stir.

Are there other universities now that are offering complementary medicine to health workers? No, although there are many courses that train doctors in complementary medicine mostly offered by doctors' associations. There are many possibilities for doctors who want to train in acupuncture, homeopathy, or in phytotherapy [*plant-based medicines*], but none of them is offered by a university. *You started a master's program at Northampton University in England.* Our program was a master's program for consciousness studies and transpersonal psychology, offered when I went there in 2005. *It seems like there's a lot of interest in psi research in England, starting in Edinburgh and spreading to universities throughout the UK.* I wouldn't call that a lot of interest and I don't think it has grown much. Most of the time it is individual researchers who are sometimes connected to universities or sometimes they are private researchers. There are a couple in Holland, in France, us in Germany, and a few people in Austria. I think there are also some in Hungary but I don't have any contact with them. There's also Etzel Cardeña at Lund University in Sweden and Patrizio Tressoldi* in Italy. Many of them are outside of universities.

Why do you think the skeptics are so eager to pounce on anything that they think isn't scientific, calling it pseudoscientific, as Wikipedia does? We are seeing a kind of religious war. There is a scientific mainstream that assumes that science has actually shown that materialism is the correct way of looking at the world. Materialism is actually a belief system that has nothing to do with science; it just happens that some scientists have looked at the world in this way. The skeptic community is a particularly convinced group of people who believe that materialism is the correct way of looking at the world. They are in a sense ideologically biased just as Islamic terrorists are ideologically biased. I'm using very strong imagery here to make the point because both groups think it's clear that what they say is true and anybody else who is of a different opinion should not only be proved incorrect, they should also be wiped off the earth, in a sense. The skeptics' community is very vocal in trying to fight these type of things because they think it is threatening a consensus of rationality which is actually not even existent. Because it's not about facts, but about the paradigms, the resistance and the fights are so cruel. Scientific fact is not just empirical evidence. It is always empirical data joined with an appropriate theory that is accepted, always sitting in the background of a world model that is also accepted. The world model that the skeptic community buys into is a materialistic world model, which they think is actually proven as true, but I think they are wrong.

You wrote the Galileo Report where you say there is not much evidence for psi phenomena in spite of much research. I said there is a lot of evidence but there is no consistent evidence that leads us to conclude that there is a phenomenon that is physically causal. This has to do with the fact that we have a lot of meta-analyses that show statistically significant strong effects. But, when it comes to identical replications that would be able to establish causality I think the skeptics are right. They say that there is no single experiment that you can give to whoever is a skilled experimenter and repeat that experiment X amount of times and always have the same result. I think we do not have that type of evidence because this is part and parcel of the signature of effects that are not causal in nature. That is different from saying there is not enough evidence. I think there is lots of evidence to say that these effects are real but they are of a different nature than getting up and switching on the light that works all the time. Sitting down and praying for the health of somebody else does not work in the same way.

What about the studies where people are shown slides and they react precognitively to the disturbing slides as an example that's been replicated in different labs. Yes, but there are also examples where these replications failed. The important thing is to see if you have a meta-analysis where you put in all the data and out comes the average effect and its significance. I would say the average effect and its significance is established but it's not established that this paradigm works every time you use it. Why that is so we do not know but I think it should be acknowledged by way of fairness and truthfulness that this is also part and parcel of the data that we have from parapsychology research. There is a lot of debate in the community about that. Some people would say we have shown it, some people say we haven't. I belong to the group that is skeptical about the replicability of these issues. Not about the generic factuality, I think that has been established, but the replicability of single paradigms is not established. I love Dean Radin's work because he is a gifted experimenter who has produced beautiful studies and lots of positive studies, but if you look carefully he always changes something. He will say he never does the same study twice because that tends to diminish the effects, it tends to bore him, and that tends to not work properly. [*I checked with Dr. Radin and he agrees he changes up experiments.*]

You teach how to do different kinds of research––whole systems observationally, qualitatively, etc. How do we accurately research psi phenomena? I'm always fond of multiple approaches. It's a little bit like agriculture. Monoculture is always a bad idea. If you have hectares of cornfields, that's boring, as are loads of experimental studies. I think what we need are experimental studies to see how do the effects show up in controlled environments. We need really good field studies to see what happens in the field and we do not have enough of that. We need good qualitative research to see how having these experiences makes a difference to people. We need a multiplicity of approaches. There is also a kind of hidden dialectics or complementarity there because the more you control things in experimental research, the less important they are for people in their real lives. The more important these effects are for people the less you can control them. The huge effects like poltergeists, NDEs, or precognitive dreams that are impactful for people, you can't control in a lab—they just happen to people. You can study them post hoc by interviewing them and then you may see huge effects. But they are out of your control; the more you can control them the smaller the effects become. Figuratively speaking, you cannot use an experiment to force people into believing. If they're open to those dimensions then the whole panoply of data will suggest to you that there is something there.

It's interesting that when people get bored with doing a repetitive task like predicting what card is going to appear the effect diminishes. Remote viewers didn't have that boredom effect so they kept on being accurate in their drawings, indicating that our emotions definitely play a part. Also the way the real world is brought into the experiment or task; if you're pressing buttons that have nothing to do with the real world it's just an instrumental abstraction and therefore the effects will be small.

What's the network that sponsored the Galileo Report? The Scientific and Medical Network is a group formed in the 70s out of mostly medical people, but also other scientists in the UK, who were uneasy about the mainstream materialistic paradigm. They wanted to open up the discussion and create a sanctuary for those who dissent to allow them to voice their opinions and discuss them with peers. They wanted to bring in evidence and people who would speak about their experiences. Out of that has grown a network of people who are more open-minded than the normal mainstream scientific community, but still see themselves as scientifically minded. *It seems like the thrust is we're creating a new paradigm that's non-materialistic. It's happening not just in England but wherever people do research.* It looks as if many people are coming from various directions and finding very similar results and very similar types of analysis, yes.

You are a historian of science; when did the wave start to crest to create a new paradigm that acknowledges the limits of materialism? I think the first instance is the development of quantum mechanics in the beginning of the 20th century. It is now reaching biology, medicine, and has reached psychology to some degree but it hasn't gone through all the sciences. In all branches of science you have people who are more open and who have various experiences that challenge a materialistic mainstream paradigm, but it has not voiced a coherent picture. It's more like a wave that ripples through various branches ever since the revolution of quantum theory in the beginning of the 20th century. *What about systems theory and chaos theory?* I think that belongs to parts of it, although you can be a system theoretical thinker and be perfectly happy with the materialistic worldview. I think there is nothing in system theory that challenges reductionist ontology. It challenges maybe a too crude reductionist ontology but it still allows you, for instance, to think consciousness is just an emerging property of a complex system like the brain.

You write about a methodology where we value the first-person in our experience and you talk about Roger Bacon (died in 1292) who advocated that method of scientific understanding. Yes, we see various moves in psychology and consciousness studies where we have a return to a first-person methodology. It's not a big strand but it's growing because people are discovering that you cannot just look at MRI pictures if you want to understand what people are experiencing. You also want to see the inner phenomenology, the personal experience of what those pictures are about. You can only discover this when you talk to people and ask them about their experiences. I think there is a strand of people in the newer science community who are advocating that phenomenological inner stance that was propagated by Chilean scientist Francisco Varela and others in the late 1980s and has grown ever since.

A group of people in psychology have started to understand that not just third-person views are valid but also valid first-person views from the inside. That has made its way as qualitative methodology in psychology and in social science, and is now even grounded more deeply into first-person methodology of contemplative neuroscience. People with experiences with contemplation describe the way their mind works. I think there are inroads now and they are increasing. *I wrote a book called* Ageism in Youth Studies *because I've researched global youth for a decade. I was astonished to find that in books about youth they did not talk to them. They would not quote a single young person, relying on multiple-choice surveys. That's another example of scientists getting so locked up in the numbers they don't talk to the people they're studying.*

You advocate a circular model for medical research; what's that? It's a direct challenge of the prevailing hierarchical model of evidence-based medicine, which is the most prevalent one. This stipulates that there is a hierarchy of methods, meaning that there are methods that are better and worse. The better methods are randomized controlled trials and if you combine the results of randomized controlled trials you have the true effect size. Everything below that like natural cohort studies, or single-case observational studies, or case series, are considered to be less reliable and have no use. The logical consequence is once you have data from randomized controlled studies you can throw away the rest. That's why most meta-analyses and systematic reviews only look at randomized controlled trials.

The argument we're making is there is no such thing as the best method or the only adequate method. There are only proper methods to address certain questions. If you have a question like, "Is a new drug better than placebo," then clearly a randomized controlled placebo-controlled study is the method of choice. If you want to know whether a new method that has been proven effective in a small group of people is also effective in a large population, then the method of choice is a big cohort study where you roll out that method to a large population, see whether it's effective and whether there are lots of side effects. The circular model would argue that you need different types of evidence put together to find the true usefulness of any intervention and part and parcel of that is randomized studies. But you also have other studies like observational studies that come into play. There is no use pitching the rigor of randomized studies against the ecological validity of other studies because internal and external validity––technically speaking––are incompatible concepts. You can't maximize both in one type of study. You can either maximize internal validity and by necessity you neglect external validity, or you maximize external validity and you have to neglect internal validity. The consequence of that situation is that you need different types of studies in order to get the full picture. *That's common sense but that doesn't mean it's believed in.* Often common sense is also not what is scientifically accepted.

You developed a questionnaire to measure exceptional human experiences. How did that work? It was one of those ideas I had and we cooked up a series of items with Niko Kohls, one of my master's students who later became my Ph.D. student, post-doc friend and colleague. He did a series of studies using that questionnaire, and then he validated it and used it for various studies. *What's an example of a question on that questionnaire?* We wanted to look at exceptional experiences that are spiritual on the one side but also may be destabilizing psychiatric experiences.

We include items like, "I have the experience of positive forces helping me," or "Sometimes I've had the experience of light helping me." Also, "My worldview breaks down," or "I had the experience that all I believe crashed," or "Sometimes I hear voices that make fun of me and chide me." It's a mix of positive and negative spiritual experiences and psychopathological experiences.

In that questionnaire we show there are four dimensions: positive spiritual experiences, negative spiritual experiences or deconstructive experiences, visionary states (a smaller factor), and psychiatric symptoms. Traditional psychiatrists often lump spiritual experiences together with hallucinations. I think that's a mistake because they are different. We can show using the questions that they are not only phenomenologically different but also psychometrically different. We translated it in English and did most of the studies in English. It's called the Exceptional Human Experiences Questionnaire.

You helped develop the Freiburg Mindfulness Inventory about mindfulness research that has to do with meditation. What are you measuring? It was developed by a student of mine who I supervised and it's a set of items to measure what we think is a mindful state of mind. It's like measuring a construct, like measuring depression. To some extent you can always do that with a self-report questionnaire and the question is, how useful is that? I think it is useful to some degree, for research purposes or if you want to track a group of people's progress through meditation programs. *It's interesting that meditation changes the brain in a physiologically measurable way.*

You are interested in homeopathy and its efficacy. It may not contain even an atom of the original calcium or phosphorus or whatever it was in the homeopathy sugar pill. How does it work? The database of homeopathy is very similar to the database of parapsychology. That led me to the conclusion that homeopathy is probably similar to parapsychology; that's why I developed the generalized entanglement model in the first place. I think we also have correlational effects here that are truly specific so it does make a difference whether you choose the right remedy or not. But it's not a stable cause like a light switch that you can switch on but some correlational effect, which we don't understand yet. There is no information, no causal effect in those remedies, but they do work nevertheless quite specifically. Sometimes I call it modern magic but that's a word that is also not very well received because it has the wrong connotation. *Dean Radin uses it...*

Some people suggest that homeopathy is vibrational, that percussing the original substance changes the vibration. It seems water can be changed like Lourdes shrine water or Masaru Emoto's water crystals.[1] These are all nice metaphors but we don't know what it is. I think we know from the empirical signature of the effects that it's very similar to what we see in parapsychology. I don't think there is any causal mechanism buried in those remedies but nevertheless the way they are produced is necessary for the effects.

What about acupuncture? That's more specific in terms of working with energy channels that can be measured and viewed. We are further down the road there. It looks like acupuncture can be conceptualized using the interstitial tissue. Research by Helene Langevin [*director of the US National Center of Complementary and Integrative Health*] showed that the connective tissue is very important for acupuncture. There might be additional tissues and additional lines of information in the connective tissues. For instance, the network of osteoblasts in the connective tissue form a true network of information, which is often neglected. It could be the case that this network of information within the connective tissue forms the basis of acupuncture meridians. There you have true physical interventions of the needle that goes into the skin. It's being pulled around, which has effects on connective tissues. Distant effects can be explained by using the network of cells within the connective tissue and the effects they have on distant electric properties such as through piezo-electric effects (electricity released through pressure on organic structures such as a dense network of molecules or cells).

People are interested in energy psychology, acupressure tapping on meridians, such as Emotional Freedom Technique that can be used by non-professional therapists. Any ideas about why they work because there is no needle? We did a study with a Ph.D. student who used Quantum Healing [*Deepak Chopra wrote a book about it*], which is an advanced energy technique. It works without any tapping at all, just touching a part of the body and intensely imagining a future state of health. It has huge effects. What I think is happening there is consciousness intervening. All these rituals are probably important for directing the concentrative aspect of conscious. *It sets your intention if you're visualizing.* It keeps you busy and active; maybe if you could do that without any ritual that's fine as well.

You edit a book series titled Neuroscience, Consciousness and Spirituality. *Where is that headed?* We publish the book series with Springer and the idea is to allow people working at the interface between neuroscience, consciousness studies, and spirituality to publish their results. We published a couple of books in that series now with more in the review process. It's an opportunity to publish work that would have difficulties in the classical journals. *The trouble with Springer is that their books are really expensive.* Yes, that is true and there is nothing we can do because that's the policy of the publishers. We did not know that before we started that series as we were a little bit naïve at that point. *Do you have another book in mind?* Yes, there are two. One is about science beyond a materialist worldview. There is probably a more extensive one, which has the working title *The Inescapability of Religion, Even for Scientists. Because materialism is a kind of religion.*

As you look at the world situation, are you optimistic or pessimistic? I'm always optimistic but I do see the grave challenges we are facing. I think we will only be able to turn the wheel around if we work together and if people are willing to make the necessary choices like living sustainably. *Germany is a leader in encouraging solar panels and wind and they make it so individuals can get involved by putting up panels, etc.* But we could do better.

As a person who has written over 150 peer-reviewed articles, book chapters and books, what do you do for fun and renewing yourself? I meditate, I sing, I meet with people. I like reading historical novels, occasionally watching good films, and going to concerts. I live in Berlin where there is a lot going on so I can go to all sorts of concerts.

Is there anything else that should be included in your chapter? It's important to install a global culture of consciousness in the sense of mental hygiene, everybody making it part of their routine to take half an hour out to clean their spirit and their mind. The image that I use is the most important invention that saved the most lives was hygiene. Physical hygiene is very natural to us as when we go to the bathroom and wash our hands, and separate toilet and drinking water. That saved a lot of lives and is still saving lives in poor countries where it's being introduced. In the same sense, we need to introduce mental hygiene. I think we need to spend as much time on clearing our mind each day in order to get rid of whatever is preventing us from rightful and useful choices in our lives.

You've been interested in teaching these kind of techniques to children, like David Lorimer in Scotland.* We tried to install a program in schools and that was quite successful but we were not able to follow that on for various reasons. Lack of funding for one, but also it's difficult in the German school system because it's very much a top-down system. You need to have access to the top decision-makers in the Ministries but we were not able to access them although it does work. We published a meta-analysis of the extant research that showed that mindfulness interventions with children improve cognitive capabilities. I would hope that more research is done and educators are creative about how to introduce that in schools. I'm teaching mindfulness to the medical students at Poznan Medical University where I teach and they are quite grateful. I teach them various techniques: body scans, breathing meditation, compassion meditation, and imagination techniques so individuals can choose whatever is most helpful for them. I think people are in different stages in their lives and need different methods and there is no such thing as the right method. There is a good method for a person in a certain situation in their life.

Books in English

The Galileo Commission Report can be ordered from the Scientific and Medical Network:
https://explore.scimednet.org/index.php/buy-the-galileo-commission-report/
https://www.galileocommission.org/

(Ed. with George Lewith and Wayne Jonas) *Clinical Research in Complementary Therapies*, 2011

Secular Spirituality: The Next Step Towards Enlightenment (Vol. 4), 2015

(Ed. with Stefan Schmidt and Wayne Jonas) *Neuroscience, Consciousness and Spirituality*, 2013, 2014, 2016

Endnote

1 https://www.masaru-emoto.net/en/science-of-messages-from-water/

Steven Taylor, Ph.D.

Beyond Materialism: A Panspiritist Perspective

Photo by John Wood

Questions to Ponder

Dr. Taylor's view of reality is expressed as "panspiritism" as an alternative to materialism. Explain.

What patterns does he find that lead to an "awakened" experience in evolved people? What's the opposite, a reverse Bodhisattva?

What does transpersonal psychology study and how does it relate to Consciousness Studies?

What might have caused the change from egalitarian to patriarchal human societies?

I was born in Manchester, England in 1967, the second of two boys. Astrology isn't really my cup of tea, but I was born on April 30th, which makes me a Taurus. I'm quite stubborn, which

I've heard is a characteristic of Taureans. But I've also read that Taureans can be quite interested in material things and are practical, kind of earthbound, which doesn't apply to me at all. I'm not materialistic; I'm not particularly practical; I'm not interested in comforts. On the Myers-Briggs, I'm an INFP, intuitive, and quite creative. I write poetry and I used to write fiction.

Do you find national characteristics as you travel around the world? I've done talks in the US, Canada, Brazil, and Europe and have found that American audiences tend to be a little more outgoing and humorous, so I tend to be a bit more lively and humorous when I do talks in the US. The national trait of English people is slightly reserved and audiences are slightly reserved as well. *I think of Brits as having a dry sense of humor.* Yes, it is quite ironic. Our humor is a bit sarcastic and quite subtle. I lived in Germany for four years and I was very aware of the slight, subtle differences in the national character, particularly in relation to humor. If I was ironic to my German friends they would usually think I was being serious. At that time I was a musician in my early 20s, playing the bass guitar and singing. My band did a tour of Germany; I met a girl, fell in love and stayed in Germany for four years because of that. When we broke up, I came back to England.

What led you to go to graduate school to become a psychologist? I was always interested in psychology and philosophy but earlier I was put off by the materialistic assumptions of some scientists—for example, the skepticism towards psychic phenomena and higher states of consciousness. I was put off by the assumption that our normal experience of the world is correct and that anything that deviates from it is aberrational. Because of that, I never saw myself as a scientist or psychologist, until I discovered the field of transpersonal psychology, through reading a book by Ken Wilber. As soon as I read about transpersonal psychology, I thought, "This is me. This is where I want to be and this is what

I want to do!" I realized that it was possible to study spiritual experiences and transformational experiences in a scientific way, from a psychological perspective.

How do you define transpersonal psychology? It literally means beyond the self, or beyond the ego. It's the study of states of consciousness beyond the ego, the study of higher states of consciousness, or spiritual experiences. I think of it as the area where Eastern philosophy meets Western psychology. It explores the overlap and common ground between psychology and Eastern philosophy and investigates spiritual traditions and practices.

In contrast, it seems that psychiatry today depends on pharmaceuticals. Particularly in the UK, I think there's a movement away from that kind of pharmacological neuroscientific perspective. A growing number of psychiatrists and psychologists are becoming interested in more holistic perspectives, such as mindfulness and ecotherapy. Some health trusts recommend courses of gardening for depressed people or encounters with nature as a therapeutic tool. I recently saw a video of our former Prime Minister Theresa May discussing mindfulness in the House of Parliament. (Unfortunately I don't think she practiced it herself!) The government is piloting programs in mindfulness in schools, so it's hopefully going to be part of the school curriculum in the UK.

In the US it seems that the insurance industry propels doctors of all kinds to work quickly. With only 15 minutes with the patients, you're not going to get to their core issues. I think people are realizing that the quick fix approach doesn't work. Research suggests that antidepressants are only effective with people who have severe depression. But they are mostly prescribed for people with mild depression, where they are usually ineffective and have serious side effects. Antidepressants don't address the core reasons why people are depressed either. A person might feel depressed because of a lack of meaning or purpose, bad relationships, an unfulfilling job, a lack of contact with nature, a lack of exercise, and so on. There should be more holistic, deep-rooted types of therapy. Mindfulness teachers like Jon Kabat-Zinn in the US and Marc Williams in the UK have been very astute because they have taken mindfulness out of its original Buddhist context and introduced it into Western culture. Some people criticize it as McMindfulness, a fast-food version of meditation but it's surely a positive sign that mindfulness has become part of the mainstream.

Before you went to Germany, did you finish your undergraduate degree? Yes, I studied literature, because I loved reading and writing poetry. I still write poetry now. Then, almost 20 years later, I did my master's degree in Consciousness Studies and Transpersonal Psychology at Liverpool John Moores University and a Ph.D. in transpersonal psychology.

You've developed a master's program like that at your university? Yes, at Leeds Beckett University we have a master's degree in Interdisciplinary Psychology. Around half the content is spiritual psychology or transpersonal psychology. The main area of my academic life is researching transformational experiences so I supervise Ph.D. students who are investigating the effects of spiritual awakening or its causes. I encourage academics to engage with the public and disseminate ideas into popular discourse. *I've heard that goal from other British professors like Chris Roe.**

What other universities offer that kind of a course? Liverpool John Moores University offers an online master's degree in Consciousness, Spirituality, and Transpersonal Psychology. (I teach a module in that degree.) There used to be a similar degree at the University of Northampton. That's no longer running, but a number of universities have modules on transpersonal psychology at the undergraduate level. I'm chair of the Transpersonal Section of the British Psychological Society, so I'm hoping to build up interest in the field.

What in your childhood influenced you to be such an achiever as a poet, a musician, a professor, a prolific writer, and a best seller? Maybe reincarnation! There was nothing in my environment that pointed me in that direction. My upbringing was very ordinary. My father was obsessed with football and sport in general and going to pubs and socializing. We had no culture-- no books, music, or theater. But when I was about 15, I changed quite significantly. I became very introverted and started to read and write poetry, essays, and stories. I began to have spiritual awakening experiences, moments of ecstatic wonder and union with my surroundings, feelings of euphoria, moments in which the world became incredibly real and seemed to make sense. Because the experiences were so different from my upbringing I wasn't sure what to make of them: I thought I was crazy. It took a few years before I began to understand myself and make sense of the experiences.

Did you have a teacher that encouraged you, because usually, students who excel in difficult environments have a mentor of some kind? No, I didn't meet any adults who were interested in the things I was interested in, like spirituality, poetry, or philosophy. But I did have a good friend called Joe, who was a year older than me. When we were about 19, we started to read the same books and realized that we were both interested in philosophy and spirituality. I thought of Joe as my guru. He was very stable and calm, very quiet and at ease with himself. He never cared about what people thought of him. I gained a lot from my friend Joe. *I can see you in a past life as a contemplative scholarly monk studying and thinking.* You might be right. The kind of person I am is strangely independent of my environment and my family. One way to interpret it would be in terms of reincarnation.

What were the triggers for those ecstatic experiences? Contact with nature, when I was walking in a park or through the fields at my school. Sometimes just walking down the street at night and looking at the sky. In my research into transformational experiences, I found that they often come in the midst of psychological turmoil and in states of depression, stress, bereavement, addiction, and so on. I was pretty unhappy at the age of 16 and 17, quite depressed and frustrated. I felt like a misfit. I didn't understand myself at all, so maybe that kind of turmoil was part of it.

Mystics typically have the dark night of the soul before their enlightenment experiences, so it seems to be a historic pattern. There is such a close relationship between suffering and spiritual experience. I don't think it's possible to achieve a deep state of wakefulness without a period of intense suffering. Suffering seems to deepen us and open us up inside. It seems to create a space within which a fuller experience of enlightenment can manifest itself. *In* The Prophet *by Kahlil Gibran, he said that suffering carves out a deeper receptacle for joy and bliss.* I think that is very true.

Bernardo Kastrup suggests that those kinds of experiences, or drugs, or deprivation reduces the filter of the brain, which allows the transcendent mind to be heard, although he's not keen on the word "filter."* I think of it more in psychological terms. When you go through intense suffering, it breaks down your normal identity. The normal ego is largely a collection of psychological attachments––to the future, past, possessions, status, ambitions, beliefs, and so on. These attachments are the building blocks of the ego. When you go through intense suffering, these building blocks are taken away––your hopes, your possessions, your status, etc. The ego itself collapses, the way a house collapses when the bricks are taken away. That can be a state of psychosis, or a breakdown. But in some people a latent higher self seems to emerge, like a butterfly emerging from a chrysalis. *So, it's not the filter of the brain, it's the filter of the ego?* Yes, you could say that.

If we were so open to Consciousness with a capital C, we might just sit in bliss and not chop wood and carry water? I've studied many cases of people who have permanently woken up, usually through intense suffering. They don't usually become incapable of functioning; they remain practical, able to wash the dishes and drive cars, and so on. But when they don't need to be focused on practical things, they can let go and return to a state of oneness. It's always there in the background.

Some people awaken, not through hardship but through spiritual practices, like meditation or sitting in silence at a retreat, or being in nature. In my book *The Leap*, I say there are three ways that people can attain wakefulness. The first way is when people are naturally awake. It's just their normal state. Young children are naturally awake in some ways. It would be wrong to say that they're enlightened, but they certainly have qualities of spiritual wakefulness. Most of us lose that as we become adults, but some people seem to retain it and it becomes more intense and more integrated. Those people often become poets, like Walt Whitman, the archetypal awakened poet, or William Wordsworth, or Mary Oliver. Sometimes they become social activists because they feel such a strong sense of compassion. *A strong sense of compassion in creatives was identified by Paul Ray and Sherry Ruth Anderson.*

The second way that people can attain wakefulness is through spiritual practices and paths, such as Buddhism, Kabbalah, or Taoism. You can't meditate for ten years without undergoing some degree of gradual spiritual awakening. The third way is when people have sudden and dramatic awakenings, usually as a result of intense psychological turmoil, such as bereavement or a diagnosis of cancer. In my research, I've tended to focus on people who have these dramatic sudden awakenings. But I think the majority of people who attain wakefulness do so through following gradual paths.

With three children, teaching and writing, how much of the time are you able to stay in that awakened state? It's difficult to be conscious of it when you're getting your kids ready for school or stuck in a traffic jam or marking students' psychology essays. In those moments it seems to be in the background. But there are always some quiet moments during the day when I'm able to tune into it. Sometimes it takes a few moments of reflection, focusing on my breath, or just closing my eyes. Maybe a few moments of meditation and then I'm usually able to tune back into it. *What's your meditation practice?* I have a few. I learned Transcendental Meditation when I was 19. I occasionally go back to that, but I also studied tantric yoga for a couple of years. I had a spiritual teacher who taught me a technique as well. I interchange between different techniques, depending on my mental state or the situation.

Are you able to teach your children these skills? I don't think you need to. I have three boys, 15, 12, and 9 years old. Young children have a natural spirituality. I don't think they need to meditate because they have the wonderful ability to be present with a vibrant energy and a natural exhilaration in being alive. As they get a bit older, they start to lose that natural wakefulness. I think by the time kids reach 16 or 17 years, it's a good time to start to meditate. Nature is a good source of a spiritual experience so we make sure our kids spend a lot of time in the countryside or natural surroundings.

In the US teenagers are increasingly anxious and depressed, especially girls. They have so many pressures to do well in school and they know they're going to have these huge debts when they graduate. They must get good jobs to pay for their college tuition and feel they must look good on social media. Is that true in the UK? I think so. There's the same pressure with exams and social media. In the UK, we have taken on the US style of education, with a lot of testing, which is quite detrimental to children's well-being. I say to my kids: it doesn't matter if you fail the test. It's not that important. *What about the A levels and O levels, which are extremely important for getting into university?* That's true, but these days in the UK you can get into a university with pretty average grades, so getting good results in A levels is not so important. *Is your 15-year-old feeling any stress?* No, he's a very chilled-out guy who doesn't worry about anything.

What have been the most difficult challenges in your life and how did you cope? You mentioned your adolescent angst. That was the most difficult time, which lasted for about 10 years beyond my adolescence. I only began to feel comfortable with myself around the age of 29 or 30. It was especially difficult when I left Germany and came back to England to live with my parents. I felt as though I had to start from scratch. I had given up music, broken up with my girlfriend, had no career and no idea of what I was going to do with my life. I felt broken down because everything had been taken away--my girlfriend, my band, my job. But another part of me was sure that everything would be fine. And within a year or so, things did start to work out. I met my wife and started to follow a spiritual lifestyle. I started to meditate regularly, quit smoking, pretty much stopped drinking, and became a vegetarian. I tried to rebuild my life in alignment with the spiritual side of my nature. And later, I decided to study transpersonal psychology. From that point on there hasn't been any struggle, although having kids is a whole set of challenges in itself!

Have you had experiences with synchronicity, precognition, as part of being in touch with the flow? Definitely. I've always believed in psi phenomena because I occasionally had experiences of them. When things are going well and I'm in alignment with my spiritual self, synchronicities do occur. When I'm writing a book, things seem to fall into place very easily.

You've written 12 books translated in 19 languages and written articles in over 50 journals and magazines. The magazine Mind, Body, Spirit *listed you as number 31 of the 100 most spiritually influential people. How do you have time to do that writing?* I'm quite a disciplined person, who can write in any circumstances like on trains, airport lounges, and airplanes. I've never had writer's block; I've never had time to! If I'm in a café, I put on music in my headphones to tune out the environment. *You must have people write to you and say your book was so transformative for them.* Yes, that happens quite a lot and I really appreciate it. I don't really think of it in personal terms. It's something that is coming through me; maybe if I became too conscious of it, it would get in the way. I feel that my role is to help the evolution of the human race in a movement towards awakening.

When I think of awakening, I think of enlightenment, Buddha sitting under the Bodhi tree until he understood how the universe operates. Do you equate awakening and enlightenment? Yes, but I don't use the word enlightenment very much, partly because it's a mistranslation of the Buddhist term *Bodhi*, whose real meaning is closer to awakening or wakefulness. Enlightenment suggests something very rarified, a kind of state that is almost impossible to reach. There are different degrees of wakefulness. Some people have a very high or intense level of wakefulness. A lot of people experience a lower intensity of wakefulness as an ongoing state. It's not a question of either being asleep or enlightened. From my research, I recognize that there are thousands of seemingly ordinary people who have attained some degree of wakefulness, usually through suffering and psychological turmoil. It's quite common and a lot of these people don't have any background in spirituality. They don't know anything about meditation or Buddhism. They just know that they feel like different people, more connected and more appreciative. They feel that life is much more meaningful. That's why I prefer the term *wakefulness* over enlightenment.

Where do you find your case studies? Do they contact you because of your books? Sometimes, but usually, I advertise my research and ask people to contact me. For example, two years ago, a colleague and I decided we wanted to do research on transformational experiences in relation to bereavement and advertised on social media. But sometimes I do make a conscious attempt to reach certain people. I'm interested in awakening experiences that occur in extraordinary circumstances and one area is combat.

I've found quite a few examples of soldiers in combat zones having awakening experiences, so I'm interested in collecting more of those. I'm also interested in the awakening experiences of incarcerated people. For instance, I've been in contact with an awakened person who's been in prison in America since he was a teenager and is now in his fifties.

In your Spiritual Science *book, you talk about a panspiritist view that transcends conventional science and religion. You said that panspiritist views explain consciousness, quantum mechanics, placebos, neuroplasticity, and evolution.* The materialist model does not explain the world very well. There are many seemingly anomalous phenomena that materialistic science simply disregards because it can't explain them, like psi and NDEs. If you replace the materialist model of science with the idea that the ground reality of the universe is a spiritual force––a fundamental consciousness or spirit––then everything begins to make much more sense. If you accept this idea, you can explain our individual human consciousness. The brain doesn't produce our consciousness, rather it's the transmitter of consciousness. Panspiritism explains the influence of the mind over the body because the mind is a more subtle expression of consciousness than the body.

This relates to the placebo effect and the physiological changes that can take place under hypnosis. Telepathy begins to make sense because there's no separation. We're all expressions of the same consciousness and therefore we can tune into each other's thoughts and intentions. Altruism makes sense, too, because we are interconnected. We can sense each other's suffering and therefore have an impulse to alleviate others' suffering. Because the whole world is an expression of this fundamental consciousness, we may sometimes feel a sense of oneness with the world, or with nature. All things *are* literally one and we sometimes sense this in deep states of meditation, NDEs, or intense spiritual experiences. The idea of some form of afterlife becomes feasible as well, because our consciousness is not directly produced by the brain. As soon as you posit the idea of a fundamental essence of spirit or consciousness that is everywhere and in everything, then you can really begin to explain the world and human behavior.

How does it explain that we're destroying the planet, inequality is increasing, autocrats are increasing, etc.? The problem is that the ego obstructs fundamental oneness. My book *The Fall* was partly a study of indigenous and prehistoric cultures that didn't have the same strong sense of ego that modern human beings have. They didn't have our sense of being separate entities living inside their bodies in separation from the environment and other individuals. They had a sense of connection to nature, to their communities and to their own bodies. Historically, a change occurred when human beings developed an intense sense of individuality and lost that sense of connection. *Karl Marx would say it happened with private property and agriculture.* I would turn it the other way around in that private property came into existence because of this psychological change, along with male domination, warfare, theistic religion, and so on. There's very little evidence of warfare until about 4000 BCE, when it started to become endemic. All these cultural changes were caused by this psychological change, this new intensified ego and sense of individuality.

Was there an impetus? Were there technological changes as Alvin Toffler suggests? Initially, it seemed to affect certain groups in certain parts of the world, particularly Central Asia and the Middle East, around 4000 BCE. These cultures became warlike, patriarchal, and developed the first polytheistic religions. One theory is that it was connected to a major environmental change that occurred in these areas. They had previously been very fertile but around 4000 BCE there was a process of desiccation and desertification. Over a few centuries these areas became very arid and life became very hard. So perhaps people had to develop a new sense of individuality to survive in response to hardship. *So it was a survival of the fittest kind of Darwinist adaptation?* Possibly, although not necessarily through natural selection and random mutations. (We'll touch on that shortly.) It's significant that the negative cultural changes occurred at the same time as many technological innovations, which helped people to survive. The changes demanded a new kind of practical intelligence, with logical and abstract thought, which was probably associated with the ego. That new practicality and intelligence was the positive side of the intensified ego.

Scholars like Marija Gimbutas said that the matriarchal cultures were peace-loving, cooperative, and egalitarian, while other scholars have said that there is no evidence of that. What did you find about the early matriarchal cultures? Warfare is definitely a late development in human history, connected with agriculture, farming, and the first large settlements. Some people call earlier societies matriarchal because they were egalitarian, but I think this is probably wrong. We shouldn't think in terms of patriarchy or matriarchy. Earlier societies were egalitarian, with no status difference or hierarchy between different genders, as in ancient Crete. There's a lot of evidence that before 2000 BCE, ancient Crete didn't have warfare and women had a high status. If you go to Crete and visit Knossos, or the museum in Heraklion, you'll see an amazing vibrant atmosphere in the artwork, conveying the sense that people revered nature with many images of beautiful natural phenomena. It's completely different from later cultures where we see a worship of warfare in artwork. *Like Zeus and Thor.* Yes, in later art, nature seems to disappear, replaced with brutal images of warfare. It appears that some kind of major psychological change occurred.

Even today Pygmy bands in Africa share in holding the babies and cooperate in net hunting. There aren't hierarchies, except older people are respected for having more wisdom. The standard view of evolution is full of fallacies, such as that human existence has always been a savage, brutal competition for survival that led us to be individualistic, competitive, materialistic, and greedy. It doesn't fit with the archaeological or anthropological evidence. Simple hunter-gatherer communities like the Pygmy bands you describe, which is how we lived for most of our time on this planet, are very egalitarian and cooperative. They didn't have private property, just like Marx said about primitive communism. A stereotype is that human groups fought each other over territory and resources but that's not true either. Different tribes cultivated ties with each other and tended to share territory and goods, with few conflicts. It's a fallacy based on modern society, a projection through the lens of modern capitalist economies.

In your book The Fall, *you suggest that social and gender inequality are a pathology.* In archaeological terms, there's a lot of evidence that prehistoric and indigenous cultures were generally egalitarian. The first hierarchal societies came into being around 4000 BCE, as a result of the intensified sense of ego in certain groups. These groups, largely consisting of people who spoke Indo-European languages, were very warlike and had very advanced technologies. So they found it easy to conquer the old egalitarian cultures and spread their patriarchal, technologically advanced societies around the world. You can tell how far they spread by the number of people around the world who speak Indo-European languages.

What interested me in academia is the similarity to Chimpanzee Politics *described by Frans de Waal but you're saying traditional indigenous people didn't have those kinds of hierarchies? And bonobos are matriarchal and solve problems with sex not fighting.* The aggression and hierarchical nature of chimpanzee communities have been exaggerated. Many studies show that chimpanzees become especially hierarchical and aggressive when their living space is disrupted or disturbed, usually by anthropologists, or when their feeding patterns are disrupted. This is just like when human communities become disturbed through overcrowding and lack of contact with nature. When chimp communities are undisturbed and natural, they have a much lower level of violence and hierarchy. Bonobos are just as closely related to human beings as chimpanzees. They point towards a fundamental egalitarianism and peacefulness and altruism. *That shows how biased science is that we focus on the chimps and not the bonobos.* Exactly.

You argue against the theory that evolution happens purely through random mutations. Are you implying an intelligently guided evolution? I think there's an impetus in evolution to move towards greater complexity and more intense consciousness. Living systems naturally move towards greater complexity. And life forms have a dynamic creativity that allows them to respond to environmental challenges. You can see this in the phenomenon of adaptive mutation, which is an accepted concept in biology now. In certain situations, mutations occur in a non-random way as needed, in response to environmental challenges. *Bruce Lipton did an experiment with yeast, which can rapidly change its DNA in order to use the available food source.* Yes, if you place bacteria that aren't able to process lactose in a lactose-rich solution, within a short amount of time they will mutate so that they become able to process lactose. This happens way too rapidly for it to occur in a random way.

That implies a creative or conscious ground of being and ties into new thinking about epigenetics, that our environment changes how our genes express. Yes, a lot of biologists believe that the standard of Neo-Darwinist evolution is too simplistic. You can't explain evolution in terms of random genetic mutations and natural selection. Biologists are becoming more aware of important factors like epigenetics, symbiogenesis (meaning that life forms cooperate in evolution), and adaptive mutation. A "Third Way in Evolution" movement aims to develop an alternative both to creationism and Neo-Darwinism.

Darwin admitted towards the end of his life that he had placed too much emphasis on the notion of natural selection and not enough on cooperation and altruism. My interpretation is that fundamental consciousness––or universal spirit, as you could call it––has a creative, dynamic quality. The universe came into being as a result of this dynamic creative potential, and that dynamism impels evolution to move towards greater complexity and more intense consciousness. I think the urge to follow spiritual paths and practices is an expression of the same evolutionary impulse to expand and intensify our consciousness. We're doing the same thing that the universe has been doing for hundreds of millions of years. That's why it feels so right! [*Jude Currivan* and Brian Josephson* agree.*]

That's a huge paradigm shift so it's no wonder that there is a lot of resistance from people who have the materialist paradigm drummed into them from school, etc. Do you experience resistance from people who say, "Steve Taylor's way out there"? Not really. I do have a strong feeling that the paradigm is shifting. In terms of my writings on psi phenomena, I thought I might get some resistance from my academic colleagues and from skeptics, but I haven't. I've actually had quite a lot of support. The *American Psychologist* journal published a very positive review of psi phenomena by Professor Etzel Cardeña of Lund University. That never would have happened 15 years ago. Obviously there are still a lot of skeptics around, but I think an opening is slowly taking place.

The municipalist movement, headquartered in Barcelona, is a growing global movement for cities to take leadership, including what they call the feminization of politics. This means cooperation rather than competition, being humble and asking the people in the neighborhood what they think. It seems the evolution you're suggesting is happening in cities. We are living through a very interesting time in which nationalism, individualism, and materialism are increasing. But there's also a strong movement in the other direction, towards increasing spirituality and cooperation, and away from mechanistic perspectives. It's not coincidental that these two extremes are manifesting at the same time. Patriarchal values are trying to assert themselves more strongly because they are feeling threatened. *I think the autocrats represent a shadow bringing old oppressions to the surface so that we can clear them.* The old separate patriarchal ego is still strong and feels threatened because of the process of awakening, with increasing empathy and cooperation.

The young women of color in the US Congress who are causing so many waves are a perfect contrast to the old patriarchal domination. Donald Trump and the people around him are very deeply asleep––individualistic, lacking in empathy, and completely ego-centered. It's strange that the process of collective awakening is happening at the same time that people who are so deeply asleep are becoming more powerful.

Do you find that happening in UK politics? At the moment we have Boris Johnson as prime minister, who is quite similar to Donald Trump. He's also surrounding himself with ruthless, non-empathic people who don't care about anything apart from their own interests. But I like to think of people like Trump and Johnson as "reverse Bodhisattvas." Bodhisattvas are normally compassionate beings who help others to awaken. Reverse Bodhisattvas are evil beings who create misery and suffering but in doing so, they also help people to awaken. Suffering is a great spur to spiritual awakening, as we've discussed.

How does that relate to your statement that, like Carlos Castaneda's Don Juan, we should consider death throughout our lives? I've found through my research that intense encounters with death are one of the major sources of transformation. When people have a serious illness like cancer or a heart attack, it can be very liberating and very transforming, bringing about a shift into wakefulness. When you become aware of the reality of death, everything changes. You realize that life is temporary, fragile, and precious. Some things become more meaningful and others become meaningless. Things like creativity, love, altruism, self-development, and spirituality become much more important. Things like caring about your appearance, keeping up with the neighbors, or promotion in your job are not so important anymore. So awareness of death can have a liberating, awakening effect.

You also are interested in time and how to change how we relate to it. Partly that comes from my interest in higher states of consciousness because one of their characteristics is a shift in time perception. Sometimes people feel they step outside of linear time and time seems to expand and open up. The past and the future collapse into the present moment, resulting in what mystics have called an "eternal now." The whole of the past and the future seem to be contained in the present. This can also happen in accidents or other emergency situations, when time seems to slow down. It suggests that time is a construct of the normal sense of self. When the normal self-system dissolves away, or is in abeyance, we step outside linear time.

Physicists know that time isn't linear and psi experience shows that you can influence the past or read into the future, so our whole concept of a linear time isn't accurate. Even in standard physics there is a notion of time as being like space, spread out in every direction at the same time. The past, present, and future are like an ocean or a panorama. And certainly in quantum physics the notion of linear time breaks down altogether. Some physicists say that in the quantum world cause and effect are reversed so that the effects can occur *after* their cause. The Transactional Interpretation of quantum physics suggests that what happens in the present is a result of waves of the future and the past coalescing in the present [*see Ruth Kastner**]. Again, the idea of linear time doesn't really apply.

The idea of quantum entanglement, that if you change the spin of one entangled photon, the other one at a distance instantaneously changes, doesn't fit with linear time either. No, those entangled effects occur across distance and time. An experiment in 2017 in China showed that entanglement occurred between one particle in a space station hundreds of kilometers above the earth and another particle on the surface of the earth. The researchers also managed to teleport one particle into the other. So entanglement shows that the standard ideas of time and space and causality are too simplistic. *They didn't just change the spin of the particle, they moved it up to the space station?* They transferred the qualities of the particles so that the one effectively became the other.

You write about the psychology of happiness. Does that imply that we can influence our state of happiness and become happier? Definitely. I teach a module on positive psychology at my university. We have control over our moods and our state of mind. For example, there is research showing that a sense of gratitude is the single most important aspect of overall well-being. Once you cultivate gratitude, it spreads into every area of your life. I sometimes ask my students to write an "appreciation list" made up of all the things they feel grateful for. I ask them to put it up somewhere in their house--on the kitchen wall, or even in the restroom--and spend some time each day reading and contemplating it. They always say that it has a powerful positive effect [as described in research and books by Professor Robert Emmons].

In terms of how you stay balanced, do you still play music? I have my guitar right next to me now. I've got a piano downstairs and an electronic drum set as well. I love to relax by picking up my guitar and finding little melodies to play. I often tune the lowest string down to a D to create a drone and improvise on top of it. It's a great way to relax, quieten my mind, open myself to creativity and allow music to come through me.

It sounds like you're optimistic about our survival as a species, even in the face of natural disasters and political disasters. Would you say that you're optimistic or pessimistic? Both. We're living in very difficult times, and certainly in terms of global warming, some of the predictions of scientists are frightening. I'm not certain that we will overcome these difficulties. On the other hand, I am optimistic because I know that a process of awakening is taking place within the collective human psyche. More and more people are opening up to spirituality. More and more people are having awakening experiences and experiencing a permanent transformation triggered by psychological turmoil. It's really a question of how quickly we can collectively wake up to save ourselves from catastrophe.

I'm trying to promote the idea of panspiritism in philosophy and science. It seems clear now that we can't explain consciousness in terms of brain activity. There is no way to explain how consciousness could arise out of atoms and molecules. An alternative idea is that consciousness has always been present in molecules and atoms. Even the tiniest atom has a tiny amount of consciousness in it. That's basically the philosophy of panpsychism. But panspiritism is different in that it says that consciousness isn't just in material particles as it's everywhere and in everything. It's the source of the universe. All material things emerge from fundamental consciousness and it pervades everything. Minds emerge in physical things when they become complex enough to be able to "canalize" fundamental consciousness into themselves. Fundamental consciousness becomes individual consciousness, via the brain. The more complex a life form is, with more brain cells and more intricate and complex connections between the cells, the more intensely it can canalize consciousness. That's why some life forms are more intensely conscious than others.

One impulse that gives rise to materialism is the desire to understand the world and have control over it. But there's so much about the world that is mysterious. *Over 95% is dark energy and dark matter that we don't really understand.* I like the phrase "over stand." If you understand something, you stand in a position of power. It's actually very liberating to let go of the desire to control and the desire to understand the world.

It means accepting the mystery and strangeness, accepting that you don't understand. As I show in my book *Spiritual Science*, panspiritism can help explain phenomena like consciousness, altruism, telepathy, the influence of the mind over the body, and so on. But there are some limits that we will never go beyond. We will never be able to overstand the universe completely.

Books

Out of Time, 2003

The Fall: The Insanity of the Ego in Human History and the Dawning of a New Era, 2005

Making Time: Why Time Seems to Pass at Different Speeds and How to Control it, 2007

Waking From Sleep: Why Awakening Experiences Occur and How to Make Them Permanent, 2010

Out of the Darkness: From Turmoil to Transformation, 2011

Back to Sanity: Healing the Madness of Our Minds, 2012

The Meaning: Poetic and Spiritual Reflections, 2013

(Ed.) *Not I, Not other than I: The Life and Teachings of Russel Williams*, 2015

The Calm Center: Reflections and Meditations for Spiritual Awakening, 2015

The Leap: The Psychology of Spiritual Awakening, 2017

Spiritual Science: Why Science Needs Spirituality to Make Sense of the World, 2018

The Clear Light: Poetic and Spiritual Reflections, 2020

Section 2

The New Physics

Imants Barušs, Ph.D.

Logical Transcendence and Meaning Fields

Photo by Paula Rayo

Questions to Ponder

What unifies Dr. Barušs' interest in world religions, mysticism, mathematics, transpersonal psychology, and prophetic dreams?

How does he define different philosophical approaches to fundamental reality or consciousness?

What's his filter and flicker theory? Are meaning fields a new conception of reality?

I was born in England in January of 1952, an Aquarius. I'm the first of three siblings with two sisters. *Your background is Latvian?* Yes, I spoke Latvian before I spoke English. I didn't learn English until I went to kindergarten. *Are there different ways of looking at life because of where we are raised?* We emigrated to Canada when I was almost two years old. I grew up in Toronto and graduated from the University of Toronto. The Latvian diaspora was very large in Toronto with a very active community. When I was a teenager, there was a great deal of pride in being Latvian and that was a large part of my upbringing.

What was there about growing up in your family that led you to be a scientist? Nothing since from my family's side, I was led toward music. My father was self-taught and conducted two choirs, a church choir (we had a large Latvian Church in Toronto) and a men's choir (made up of WWII veterans--everybody came through the war). From the time I was a child, I was singing in his choirs. The dean of the Faculty of Music at the University of Toronto was Latvian; he knew my abilities and strongly encouraged me to go into music. But, I ended up going to one of four high schools in Toronto that had technical programs, including a Science, Technology and Trades (STT) program. I picked that program, not because I was inclined to it, but because I would be learning more things. I was very good in high school at everything, so I was naturally streamed into engineering. I received the gold medal in STT in grade 12. I also went to grade 13 which was almost like university level courses today [like the British model].

I applied to engineering science at the University of Toronto, which is a very elite engineering program. They have a program called engineering science to take material from physics and create new engineering. I received honors standing at the end of my first year, but in my second year, I had an existential meltdown because I was trying to figure out the point of life. I started reading about world religions in high school. In grade 12 history class I asked the teacher, "Why don't we learn the history of the East; why are we always learning about Europe and North America?" His answer was, "Because they didn't do anything. Why don't you make up the lecture for the last week of class and tell us why we should study the history of the East?" So, that drove me to do my research. The argument that I proposed was that in the West we had focused on a practical materialist utilitarian approach to life, while in the East they are more interested in spiritual development. The students loved it, but I don't think any of the teachers said a word to me afterwards about it.

I had very low self-esteem so I'd never thought I'd go anywhere, do anything or be anything. *Why, if you were so accomplished in high school and got gold medals?* I had a negative attributional style[1] which I still have to some extent—the feeling that there must be some mistake. *The impostor syndrome?* Yes, my father worked as a carpenter in Toronto. In Latvia he was in his last year of law school when the war forced him to terminate his study and join the army. It created my belief that you couldn't really get anywhere in life. In my second year, I questioned, "Why are we here?" I read and wrote poetry and struggled with the big issues of life. The other thing I did in my second year was audition for the Mendelssohn Choir, regarded as one of the best amateur choirs in the world and I got in. It was a very heavy schedule, plus my second year engineering and my existential crisis, and they all collided.

I dropped out of the engineering program and entered a program called New Program, where you could basically take anything you wanted. I took courses that I hoped would help me understand the point of life and answer these existential questions. I took a philosophy course, a study of Martin Heidegger's *Being and Time*, which is not easy reading. I put an enormous amount of effort into reading and trying to understand what Heidegger was saying. That created a shift for me in the sense that it became clear that most people were living out their lives in an inauthentic state. They weren't really grounded in their own understanding of what it is that they could be doing. I took more courses in existential philosophy and analytic philosophy, like "Philosophy of Mind" in upper-level courses.

Then I wondered, "But what is it that's doing the philosophizing?" so what I really wanted to know is psychology.

But in the early 1970s, is was all rats and pigeons, so I quickly realized that psychology was not the right discipline for me, except I took several courses to learn about the brain in physiological psychology, what today is called neuroscience. I felt university was for fools, that it was pointless, but as a matter of personal discipline I needed to complete a four-year degree, not because I believed in it anymore. I couldn't find anything that I felt was addressing what I wanted to learn. I tried Religious Studies but I couldn't stand the atheists teaching them, except for a full-year course in Daoism that was not taught by an atheist; I loved Daoism. From my study of world religions, I was attracted to Daoism, Hinduism and mystical Christianity, because they spoke to these deep states of consciousness and being. Mathematics had that same pull for me, the really abstract level of mathematics. I had this intuitive sense that it was similar to what was happening with people in mystical states of consciousness.

You see a connection between math and mystical states? It felt like the state where you go in mathematics is the same place you go as a mystic. *It's abstract and non-material?* Yes, you go into this abstract space where everything is glued together and connected, something deeply meaningful. I had this intuition in high school and still have that sort of feeling. I was very good at mathematics in high school and I had a teacher who gave me extra work and gave me a lot of IQ tests because he was interested in that. I wanted to learn about the underlying order to reality that's not the way we usually think reality is. I took courses in physics and mathematics to fill out the curriculum. I ended up taking, again without the proper prerequisites, third and fourth year physics courses, some mathematics courses, a course in statistics, and graduated with a four-year degree.

I was at a friend's place one night who had a roofing company, who jokingly said, "You should be a roofer." He was a high school dropout and Latvian. We had a good laugh and I went home. After my graduation, I had no job. I looked into office jobs and other jobs I couldn't see myself doing. So I spent five years as a roofer, which was a good thing in a number of ways. I'd been sitting in classrooms my whole life and I thought that I needed to be out there. Roofing is kind of a rogue trade in the sense that you're in your own world where nobody cares what you're saying, how you're saying it, what you think about--there is this freedom. You are not constrained by all this academic nonsense that I had had enough of.

I ended up moving out west working in Calgary for the biggest roofing company there. I learned to work with an enormous amount of intensity because if I started lollygagging, I wouldn't make any money. In retrospect I feel that that was very important for me because otherwise I would not have survived as a graduate student in mathematics because that's what I ended up doing next. I made a decision in 1978 that I needed to go back inside the university system and try to do what I could to help to change it. I listened to gurus, I practiced meditation techniques, I learned psychosynthesis. The latter is a theory of the psyche and system of psychotherapy developed by Roberto Assagioli in 1910. I like it a lot. It's discussed in Piero Ferrucci's book *Your Inner Will*--he was a student of Assagioli's.

I studied the Enneagram [I'm an 8] and learned how to cast a person's astrological chart.[2] It was all outside the system in the shadow culture, but I felt that's not changing the institutions of our society. We need to recognize the importance of spiritual development and acknowledgment of the need to develop meaning in one's life, which is missing from our formal institutions. They haven't changed and can't tell the difference in many cases between someone who is suffering from pathology and someone who is suffering from exceptional well-being or spiritual emergency or spiritual emergence. *Like a kundalini awakening?* Yes, Stanislav Grof, for example, started the Spiritual Emergency Network, later renamed the Spiritual

Emergence Network. Psychedelics and apparent kundalini rising experiences can both be traumatic "spiritual" awakening experiences, hence the need for such an organization. He was also involved in LSD research which morphed into the holotropic breathing method.

I decided that I needed to go back inside the system and do what it takes to succeed and so I looked through the university catalogue. I considered Religious Studies but I didn't want to learn Sanskrit or Hebrew. Not much had changed in psychology in 1978. The only thing that I could do was mathematics because you can't contaminate mathematics; you can either prove theorems or you can't. What else you think is of no substance. All the roofing gave me the intensity, the training of the will, the stamina to do mathematics. Without it, I don't believe I could have succeeded.

The other reason for studying mathematics was because I am interested in transcendent states, in meaning, in mystical states. I don't see the world through a materialist lens. This goes back to that intuition that I had back in high school that part of mathematics is mystical. I felt that mathematics was the most spiritual discipline that I could engage in. I thought, "If you succeed at mathematics, then people at least can't say that you are weak-headed." Since I am interested in where things come from, I started asking where does mathematics come from? My specialization officially is Mathematical Foundations, mostly made up of mathematical logic.

That ended up being a very difficult experience. I didn't have an undergraduate degree in mathematics. In Calgary I was part of the way up a roof when I got in my van, drove to the university and asked for the graduate student counselor. I walked into his office wearing my roofing clothes with tar on my pants and face, long hair, with a beard halfway down my chest. He said, "Why don't you take these three courses and if you can get an A in them then we'll think about it." So, I ended up taking three third-year courses. I was required to get As, which I did, and they admitted me. I worked as a research assistant and a teaching assistant for the university, and no longer as a roofer. They sent me to a professor in the philosophy department, Verena Huber-Dyson, who was actually a mathematician. Her way of teaching was, "Come in twice a week for two hours at a time and teach it to me." So, that's how I learned category theory and topos theory by teaching it to her. By the time I finished all of this, my lack of self-esteem was gone and I felt that you could show me anything in any discipline and if you give me enough time, I can figure it out. *You finished your master's in math?* Yes, it took me four years.

Then I switched to psychology because I was still interested in these mystical states of consciousness and how we find meaning in life. Four schools in California offer alternative programs: the California Institute of Integral Studies (CIIS), the Institute for Transpersonal Psychology—now Sofia University, Saybrook University, and John F. Kennedy University. They all had programs in transpersonal psychology or something having to do with consciousness and spirituality. I applied to CIIS and was accepted into their Master's Degree in Integral Counseling Psychology in 1983 and spent six months in that program. I planned to stay longer but I wasn't allowed to work with a student visa. I found that there was an alternative psychology program at the University of Regina in the Canadian Prairies, so I came back to Canada.

What I didn't know before I got there was that that whole program had collapsed. The university was considering dissolving the Psychology Department. The only reason they allowed me in was that they thought that I was a sure bet probably because the letters of recommendation from my math and CIIS professors were so strong. I could do anything I wanted and that's why I went there in the first place. I thought, "How am I going to get to study what I am really interested in, which is sort of spirituality? What if I become an expert at consciousness?"

My candidacy exams were specifically about consciousness. I found the thesis advisor, Bob Moore, at the Catholic college affiliated with the university and he was allowed to supervise me. He didn't know anything about consciousness but that was fine because if I could teach myself math, I could teach myself consciousness, so I didn't care.

I started reading everything I could find written about consciousness in philosophy, psychology and neuroscience. I wrote my Ph.D. about how people conceptualize consciousness and what that has to do with their beliefs about reality. I found that there was a material transcendental dimension; some people are materialists so they see consciousness in very restricted ways as a by-product of brain activity or computation. Dualists say that there is something else going on in addition to the physical world. We call these the "conservatively transcendent," based on the data that we gathered from the surveys carried out for my Ph.D. dissertation. The "extraordinary transcendent" people think that consciousness is a primary aspect of reality, that everything else is secondary, including matter. *Are you in the third category?* I would say I am there. My thesis was eventually published as a 1990 book, *The Personal Nature of Notions of Consciousness: A Theoretical and Empirical Examination of the Role of the Personal in the Understanding of Consciousness.* Obviously a Ph.D. dissertation, nobody would write a real book with that title. I also published 10 papers about the study.

Did you have your own personal transcendent experiences through meditation or other ways? My experiences show me that reality is not just what we think it is, such as my experiences of precognitive dreaming discussed in *The Impossible Happens: A Scientist's Personal Discovery of the Extraordinary Nature of Reality.* In the early 1970s I decided to become a vegetarian and had a dream in which I had been eating meat all along. I found out that, in those days, cheese usually contained rennet from the intestinal linings of sheep. About seven or eight years ago, I became gluten-free, but then had dreams in which I was eating some kind of white bread and trying to pull it out of my mouth. There were three or four dreams like that before I learned that couscous was actually wheat. I thought it was a grain and was eating it every morning for breakfast. Once I stopped eating it, the dreams stopped.

At the University of Regina, one of the members of my committee had me reading David Woodruff Smith and Ronald McIntyre who developed a model using classical logic for Edmund Husserl's intentionality. He is a German phenomenologist who had an idea about how conscious mental acts are structured. Smith and McIntyre show how you could apply classical logic to an analysis of intentionality. I realized right away that you could use category theory a lot more efficiently to get a much better fit. My professor suggested I write an article, which I sent to *Husserl Studies*. In Regina there was nothing to do but play hockey. It's dark in the winter, very cold, often minus 40 at night. I had a dream where we were playing floor hockey with orange plastic pucks and I took a shot from a long distance from the net and to my surprise, it went in. A week or two later, I got a letter from the editor of *Husserl Studies* raving about the paper but stating that I needed to explain the math that I was using because they had no idea what that was about. I thought, "Wow, the dream came true. I took a long shot and scored."

I started watching my dreams and noticed that I had two types that had to do with academic publication--hockey dreams or dreams about lotteries. On my first sabbatical, I was writing my second book, *Authentic Knowing*, and needed a publisher. I dreamed I had gone to the mall where there was a lottery for a black Cadillac and my friend had bought me the winning ticket. Upon awakening, I recognized who that friend was and through him, I was able to get the book published by his university's press. No publishers would pick up my book *Science As A Spiritual Practice*, but then I had a lottery dream in which

I dreamed the date on which the book would be accepted for publication and I did receive an offer on that date. It became clear that this isn't made-up stuff, that somehow our psyche has the ability to transcend time and to represent for us what is happening in symbolic representational form. When we are asleep, our guard is down and we have access to this deeper part of our psyche and can see a pattern.

How does quantum physics fit into your definition of consciousness? During one of my sabbaticals, I sat in on an intermediate quantum mechanics course and one on quantum field theory, the next step up from quantum mechanics. People like to use quantum mechanics and quantum field theory like the philosopher's stone, like you touch something with it and it magically transforms. Once you actually know the details, it's not nearly as easy to make things happen with it. I published about five or six papers on quantum mind, available on the open access website on my university's server (https://baruss.ca/pub.htm). Quantum field theory is a lot more interesting and more powerful than quantum mechanics with its presentation of particles in a field. You've lost classical mechanics' notion of continuously existing bodies moving. That's gone in quantum field theory, whereas quantum mechanics still holds on to those ways of thinking.

I like the experiments in quantum mechanics, like the Cheshire cat experiment, for example, where you send a neutron's mass down one path and its spin down a separate path. Basically, you are spatially separating the properties of a subatomic particle. There are also quantum eraser experiments, which are a lot more difficult to explain. What quantum mechanics and quantum field theory do is take down a billiard-ball version of the universe and show that that's not the way physical reality works, that it's much more interesting.

It's potentially wave or particle, but it all exists in potentiality until it's observed? That's an artificial duality because the whole idea of a particle doesn't make any sense. If you have something that has no continuous existence, then in what sense is it still a particle? And nothing is waving. It's a "probability wave" that accounts for the sort of interference patterns that you see in double-slit experiments. But it doesn't mean that whatever that was, was a wave. Any of the metaphors that we come up with for trying to talk about physical realities are going to be inadequate. The mathematics only tells you what will happen in experiments; the mathematics does not describe what that reality is, and that's very important to understand. We have mathematical formulations now, for example, by Bob Coecke at Oxford University, which are quite different from the standard sort of tensor algebra formulations that you see in quantum mechanics and quantum field theory. The mathematics presents a different kind of picture about what reality might look like so you can't go by what the mathematics looks like and you can't go by your metaphors. So, the reality is that the underlying physical reality is so enigmatic that we don't have adequate conceptual tools really for getting at it. We have to go beyond quantum mechanics and quantum field theory. We need new ways of thinking about reality.

You mentioned that nothing really happens until you observe it. In the interpretations of quantum mechanics in which there is a collapse, you have a superposition of states and then one of them gets chosen and that's the one that we live in. But there are versions of quantum mechanics that don't involve any collapse; in Relational Quantum Mechanics, there is no collapse. And basically what you end up with is all possible paths taken at all times. There is no longer an agreement on the idea that there is a collapse. If you think there is a collapse then what you are going to see in the experiment is going to depend upon what you choose to measure. With the Kochen-Specker theorem, the values you get either don't exist until you choose what to measure or they change depending on what else you are measuring, because in quantum mechanics basic measurements come in pairs. I am not exactly sure why, but if you change the other part of the pair, you'll get a different value for the first part.

I proposed a flicker theory in which the universe flickers on and off. In *Transcendent Mind,* written with Julia Mossbridge, we developed the filter theory. In the flicker and flicker-filter theories, we get at the same difference that David Bohm noted with the distinction between the Implicate and Explicate Orders. There is an inner order, we call the *deep consciousness* or *pre-physical substrate,* and what we see as physical manifestation. I no longer talk about the physical world; I talk about it as physical manifestation because it's not clear to me that there's actually anything there. Basically, everything is a manifestation of mind.

I have an interesting experience that relates to that. I was in a lucid dream where I was in a tall building in an empty white room. There were no doors or windows. I remembered Samuel Johnson's refutation of Bishop George Berkeley's 18th century idealism. Berkeley gave his version during a church service in which he presented a thesis that everything is made out of consciousness or some kind of immaterial substrate. After the service, Samuel Johnson said, "I refute him thus," and kicked a large stone to show that physical reality is physical, that matter exists. So, I remembered Samuel Johnson's refutation in my dream and thought, "Let's do the refutation." I walked over to the wall to my right and pounded on it with my fist and it was solid wall. This is what Charles Tart* calls "state-specific sciences" in his 1972 paper.[3]

I wonder about my dream wall, is that made out of molecules? Clearly not. In the same way, I doubt that physical manifestation is made out of molecules. We know that we don't know what those are. That's the very bottom of it; it's enigmatic. We don't have molecules. We have experiences that reliably play out as though there were molecules––until they don't. There's all sorts of extraneous baggage around the concept of "molecule."

Molecules themselves are made out of atoms and atoms are made out of "subatomic particles." But there is no such "thing" as a subatomic particle. So what happens to the molecules that are not made out of anything? *Because 95% is dark matter or dark energy; we don't know what that is?* Right, we don't have any idea what that is. What I am saying is everything is a product of the mind and consciousness so quantum field theories are transition theories that can help us get at a way of thinking.

Most recently I've come up with the idea of meaning fields as ways of structuring reality. Basically, the idea is that nature is not stupid, not blind and not mechanical. There is an experience of meaning, as discussed in an article published in *EdgeScience Magazine* about meaning fields.[4] In a new book, *Radical Transformation: The Unexpected Interplay of Consciousness and Reality,* I give the details of meaning fields. I also developed a mathematical model of meaning fields. That paper was published in the journal *Consciousness.*[5] [*In the magazine article the author defines meaning fields as carrying "the necessary knowledge and ability to intelligently structure events in physical manifestation." He was led to rethink the "fundamental structure of the universe" as including a trans-human knowledge and connection due to the evidence of non-local effects such as his own experiments doing remote healing using Matrix Energetics, Bill Bengston's* experiments with mice inserted with breast cancer where control mice were also healed, remote viewing, physicist Lee Smolin's discovery that similar hydrogen atoms can copy each other's properties from a distance, quantum eraser experiments, etc.*]

Do meaning fields exist apart from humans or entities? Are they more than thought forms? Is there a deep consciousness on its own? That is, if there were no humans or entities, would there be meaning fields? Are they what some would view as God, creator, universal intelligence? Depending on how you define "entity," meaning fields are entities. They are more than thought forms that I usually think of as insentient, whereas meaning fields have qualities of mind. Whether they have qualia or not I don't know, but they are capable of making decisions based on making non-algorithmic judgments. If there were no humans or entities, there would still be meaning fields. Note that the way I have defined them, meaning fields are the explicated version of Bohm's implicate meanings, so they are part of manifestation. *"God, creator, universal intelligence?"* I don't think so, but aspects of that, perhaps--something more like devas (non-human beings) that structure reality.

What difference does it make if time flickers or is constant? Flickering time means that anything can happen while time is "off." It enables manifestation to switch to something completely different. As I argued in one of my papers, within the span of Planck time, there are no discernable gradations of time, so what does it mean to say that time is "continuous"? My guess is that it is neither continuous nor discontinuous. That those qualifiers do not apply to it. There is just now, which keeps being regenerated. But those regenerations are not continuously strung together, even though we experience them that way.

So we don't really know what consciousness is. Of course not, although consciousness is an essential aspect of reality. It's primitive relative to matter certainly because if you look at all of the evidence carefully, matter is derivative upon consciousness, but we don't know what consciousness is. The consciousness that we experience in the brain is only a stepped-down or filtered form of consciousness. The meaning-making mental conscious reality that underlies physical manifestation gets stepped down as the consciousness that we experience in the brain.

Do you know that or are you just saying logically that it must be? Hindus and Buddhists have said the material world is illusion. How do you know anything? You can always be fooled. So it's a matter of the degree to which you think that something is happening is the most reasonable way of thinking. Every other track that you try doesn't fit the facts as well. *So there is some kind of universal intelligence?* Oh yes, the meaning fields are intelligent. They can make judgments. I am not sure if they have existential meaning or not. But they have inherent meaning and I think that there is a kind of intelligence that is not just a product of human conceptual work.

But why do you think that? Because the alternatives don't work; it's the most reasonable way to think about it. I've been thinking about this since I started reading books about Eastern philosophies and studied all this physics and logic. You have to take into account all of the evidence too. This is why I am always interested in anomalous data and phenomena that show you where the edges of your conceptual world lie. If there's one thing that doesn't fit, it means that you don't have the right picture of the world. For example, we know physical reality is not made out of particles anymore. There's got to be something else and the only thing that's consistent that works is that there is some kind of consciousness that is creating everything. *Can paranormal psi phenomenon be explained in terms of the meaning fields?* Meaning fields are just one element in the whole explanatory picture. They govern the regular relationships between events that you experience, which we call physical manifestation. There is no distinction between psi phenomena and non-psi phenomena. There are just rules that vary with the different meaning fields.

What's the title of your next book? Radical Transformation: The Unexpected Interplay of Consciousness and Reality. I am also writing the second edition of *Alterations of Consciousness.*

What do you do for fun? I write books. I love teaching. I love my students and love mentoring students doing research. *Do you still play hockey?* Yes, I am down to once a week. *Have you got married along the way?* Never married. I just haven't found the right person.

The world seems like it's falling apart. Are you an optimist or a pessimist? The new book's Chapter Four is about global crises. When will this slide into stupidity, self-destruction, arrogance, irresponsibility, etc. stop? I explain why I don't think you can actually stop the downward slide because that requires people actually waking up, coming to maturity, and the opportunities for doing that are being lost. So unless there is some kind of intervention by somebody who knows what they are doing. *The Pleiadeans*. Yes, I've got aliens in that chapter but there needs to be some kind of intervention. I don't know whether to be an optimist or a pessimist. You can't do anything except do the best you can and try to make things better. You really led me down this, down this… *rabbit hole?* path of self-reflection.

Books

The Personal Nature of Notions of Consciousness, 1990

Authentic Knowing: The Convergence of Science and Spiritual Aspiration, 1996

Alterations of Consciousness: An Empirical Analysis for Social Scientists, 2003

Science as a Spiritual Practice, 2007

The Impossible Happens: A Scientist's Personal Discovery of the Extraordinary Nature of Reality, 2013

Transcendent Mind: Rethinking the Science of Consciousness, with Julia Mossbridge, 2017

www.baruss.ca

Endnotes

1 https://positivepsychologyprogram.com/explanatory-styles-optimism/

2 https://www.enneagraminstitute.com/type-descriptions

3 Tart, C.T. (1972) "States of consciousness and state-specific sciences." *Science*, 176 (4038), 1203-1210.

4 https://www.scientificexploration.org/edgescience/edgescience-issue-35

5 https://digitalcommons.ciis.edu/conscjournal/vol7/iss7/1/

Jude Currivan, Ph.D.

A Conscious Holographic Universe

Photo used by permission

Questions to Ponder

What principles guide Dr. Currivan in accessing information from nonphysical beings?

How does she assist in healing our destructive worldview?

Dr. Currivan emphasizes the evidence that we live in an informed and holographic universe. How does this framework explain psi phenomena?

What's the elephant in the room and how do you define it?

I was born in Chesterfield, a small town in the north of England on the 17th of March 1952, St. Patrick's Day. I was the eldest with a younger brother. My sun is Pisces and my moon is Sagittarius, which I share with Albert Einstein. My ascendant is Leo. *Do you identify with being a Pisces?* As an astrologer as well as a cosmologist, I tend not to just look at the sun sign, because that's the outward radiance. At the very least, if I was working with someone, I would look at their sun, moon, and ascendant, because that's rather like the head, heart, and hands.[1] It's like the outer, the inner, and the path through life. I very much resonate with that sense of compassion, spirituality and inclusivity that is Piscean. My moon in Sagittarius, the inner level, influences a lifelong curiosity, a lifelong journey to understand the meaning of life: why, what, and how? My path has been one of Leo where I've been in leadership roles or I've been in more public roles.

I've got a grand square in my chart, which includes Chiron, so that sense of healing has been a very important part of my life. I have an opposition between Jupiter and Saturn. For many years, I experienced this as a push-pull tension but over time realized just how much this opposition allowed me to both express and balance the expansive out-breaths and grounding in-breaths of life. It's been an incredible support and an empowerment of my journey so I think knowing our astrological influences is very helpful and healthy for all of us. *You're an ENFJ.*

Were you raised in the Church of England, and if so, was your mum concerned about you hearing voices? No, she wasn't, because I never told her. She used to say I'd talk with anybody, but I never spoke to her about all of the discarnate "everybodies" I was communicating with. It wasn't because I had any fear and thought she'd say, "That's nonsense." I'm sure she would have absolutely been open to that. It just never occurred to me. We went to the local Methodist church but I rebelled, fairly early on. Mum was a very wise woman: Everybody in the neighborhood used to come to her when they were in trouble. When I was older, she said that she could tell my brother what to do, but she had to help me think it was my idea to do what she wanted!

What themes in your childhood led you to Oxford? There probably weren't many women when you were a student. That's for sure. My master's was in physics, so there were six women to around 200 men in my year but that never worried me. The difference for me was rather cultural in that I'm the daughter and granddaughter of a coal miner from the north of England. Most of the people who went to Oxford in those days went to private schools and came from a very different socioeconomic background than myself and most came from the south of England, so cultural differences had more impact than gender differences.

From the age of four and five, curiosity was my middle name, and that Sagittarian moon is certainly part of that. I was intensely curious about reality and the nature of the world. I started looking up into the sky and falling in love with the cosmos and the universe from a very early age, so I was studying astronomy and quantum physics from the age of five and six. Mum taught me to read at the age of three and so I was reading science books aimed at teenagers that were simple but to me inspiring. I was also "walking between worlds" having OBEs, telepathic experiences, precognitive dreams, and remote viewing and was communicated with clairaudiently by a discarnate being, who named himself Thoth and who became my guide and mentor.

My stepdad died when I was ten but my mum was and remains amazing in her support and wanted me to have every opportunity. I was the first person in my family to go to any university and because my dad had been a coal miner, I got a miner's union scholarship to Oxford. I was supported in going there financially, which made a huge difference. I continue to be curious and my work interweaves leading-edge science, across numerous fields of study with wisdom teachings from many traditions and ongoing experiential understanding of the nature of consciousness and reality.

Kids with disadvantaged backgrounds have some kind of mentor who says, "You're bright, I want you to go to a university." No, I had some very kind teachers, but nobody was championing me to do that. My champions were Thoth and my mum. When I turned 16 and started to study for A levels, I won a lot of school prizes for my work and was at the top of the class. In those days, before A levels, you could sit the admission examinations for Oxford and Cambridge. Mum pushed very hard for me to be invited to sit those exams.

Did your brother go to university as well? No, we're very different and yet we unconditionally adore each other. He went into a factory role and he's now a landscape gardener for the town he lives in and he's wonderful. He got married young and has two sons and a number of grandchildren and loves being a granddad! Our lives are very different but we're totally supportive of the paths that each has taken. Certainly some of my cultural norms are very northern. For example, if someone comes to the house, if the kettle is not on for a cup of tea within 30 seconds, I've failed. My mum left school at 14 and was not given anything like the educational opportunities I've had. She was one of a very large family, very working-class. My grandma was also a strong woman. So there's a lot of resilience in the family, emotionally and physically, which has always stood me in great stead.

What have been the most difficult challenges in your life? Losing my stepdad when I was ten. I never knew my biological dad; the person I called "Dad" adopted me and my brother when he married Mum when I was around three. Another challenge was after going through various disastrous love affairs, I sort of retreated into my head. For many years, I went around the old unhealed pattern of codependent relationships so felt a yearning and a sense of incompleteness. It took me a long time to realize that I could never fulfill that sense of wholeness through a relationship until I remembered my own wholeness. It took me well into my late 40s and through many challenges to really do that. Then the love of my life walked in and said, "Hello, here I am." We've been married 17 years. He's Curran, the hero of my novel. I'm the heroine, of course, Elanor, and he's my hero every day.

Do you have children? I lucked out, as our American friends say. Tony had two children by a previous relationship and they became my stepchildren. They are amazing and wonderful, so they are my great heroes as well. Alice is 25 and William's 27. I didn't have to do the nappy stage: I inherited them as part of the honor of being their stepmom, when they were seven and nine.

Tell us about Thoth, an Egyptian god. I was very much into ancient Egyptian wisdom teachings from age four when he first came to me. It seemed very natural because he was gentle and kind. He said to me early on, "We're not gods. We are principles of consciousness." He didn't use those words when I was four, but that was the sense, so it wasn't about worshiping anything or anyone. He is an archetype, as are many other discarnate beings, of higher level of intelligences that all of us innately have access to, if we're willing to be open to those sort of communications.

How did you first become aware of Thoth? I was in my bedroom and a discarnate light came into the corner of the room like an orb. I started hearing clairaudiently and entered into a conversation. I felt loved and accepted. Even then, I was feeling a bit like a fish out of water, so I sensed that this is where I belong, this is what I understand at a deep level without knowing why. *Have you maintained that communication with Thoth all these years?* Yes, for 63 years. Through my opening up to our innate supernormal facilities, I've learned from many other discarnate beings, levels of intelligence, and archetypes; I'm in ongoing communication with them every day. *Do they help you write your books?* I feel that they are very supportive and the books write me, in some way, as a community exercise.

If someone said to you, "I would like to have that kind of assistance and communication with beings more intelligent than me. How do you do it? I'm not aware of my guides or these archetypal intelligences that I can communicate with." One thing that he taught me very early on was, "Don't think of us as a hierarchy. Think of all of us as being part of a cosmic symphony of consciousness." That was incredibly helpful and empowering. He also advised, "Whoever you communicate with, always ask the questions, 'Does this feel of integrity? Does this get me into my ego, or does it let me be free of my ego? Does it feel resonant with my heart wisdom? And, finally, are the communications consistent?'" Those have been the guiding principles that have been with me all through my life.

Everybody is a microcosmic co-creator of the multidimensional intelligence of consciousness of our universe and beyond. The first step is just showing up and being willing to be open to the possibility of these sorts of communications. Maybe begin by being in nature and being open to the possibilities of the natural world communicating with us, through the realms of animal and plant communication.[2] Then open up to the multidimensions of the elemental realms, the devic realms and the angelic realms. When we are willing and we let our ego gently rest by our side, when we do "hear" something, whether it's some form of guidance, some form of intuitive insight, don't ignore it. Listen to it. Follow it. When we experience synchronicities, see them as everyday miracles and way-showers, not as, "That was curious, what a coincidence," and just walk on by. Take these communications as real opportunities to enter into that wonderful adventure of an exploration of multidimensional realities.

Meditation helps as well, I would think, because it can quiet the left brain, the chatter. Different things work for different folks and meditation's got thousands of years of validation for its benefits. And certainly in the West, we're very busy in our doing-ness, our thinking-ness, so anything that can help slow us down and open up that inner connection is beneficial. It could be meditation or mindfulness, yoga, dance, being in nature or finding a quiet space and taking an in-breath, because we're not very good at breathing. Just slowing down our breathing and being in the present opens up those channels of possibility, in my experience.

To go back to Oxford, did you major in physics as an undergraduate?

The way that Oxford works is, because it sees itself as a very prestigious university, you go in to take a masters as a three-year course, because the level of intensity and the level of intellectual challenge it offers is the equivalent to going beyond the undergraduate level. *You probably weren't able to talk about your visionary insights with fellow students and professors?* Correct. Again, it didn't occur to me. I went to Oxford because, having studied quantum physics and astrophysics in the years before I went, and then taking A levels in physics and mathematics and other subjects, I naively thought I would encounter the leading-edge thinking of science. I actually did; it's just that it wasn't as leading-edge as I hoped it might be.

Fortunately, I had a wonderful mentor called Dennis Sciama who was also Stephen Hawking's mentor. Dennis took me under his wing and got me invited to postgraduate seminars when I was 19 years old, one of which was taught by Hawking and where I was the youngest person in the room and I think the only woman. Dennis encouraged me to enter an essay competition, so I wrote on black holes, which at the time were becoming news in the scientific world and won the Johnson Memorial Prize for Physics. That encouragement was lovely because Oxford wasn't a terribly caring environment for students in those days. You were just thrown in and expected to swim.

How did you get from a master's in physics to being a business leader and doing innovative business work? The bridge was really the universe knocking me sideways in my third and final year at Oxford. I was in the midst of a particularly difficult relationship, and I fell apart emotionally and was not fully present so I didn't get the level of degree that I was presumed to get. I was expected to sail through into a doctorate and into academia. The universe had other ideas for me and thank goodness it did, because if I'd gone into academia, I'd have been limited and surrounded by peer-group pressure to conform. I knew from what I'd experienced at Oxford how limited mainstream science was––albeit leading-edge at the time––and has remained so to this day.

In my most recent book *The Cosmic Hologram* I write about the emergent perspective of integral reality, where consciousness and mind aren't something we have, it's what we and whole world are—and I describe the informational and holographic scientific framework and increasingly compelling evidence for this at all scales of existence. However, that came much later. In the meantime, having been helpfully knocked sideways by the universe, and knowing that I was very good at maths, I found myself entering business life and trained to be an accountant. In the next two decades I rose very rapidly to a position where, in the early 1990s, I was the most senior businesswoman in the UK. Such a career was wonderful because it made sure that I was grounded in society, organizationally and globally. That thread has carried on, even though I left business of my own volition in 1995/96, as I am asked to speak about transformational changes around the world, including for organizations and leadership. That 25 years of business experience has been extremely helpful as a foundation for what I do now.

People blame the multinational corporations for destroying the planet and the growing inequality between the 1% and the 99%. Do you have any hope that business is becoming conscientious, or is it still focused on the short-term bottom line, getting a profit to the shareholders? As a healer, as well as my other hats, I know that a healing journey doesn't begin until we acknowledge and take responsibility for our own dis-ease. Our dysfunctional behaviors, whether it's on a personal scale, organizationally, nationally or collectively, are consequences of our dis-ease of awareness. We have fragmented perspectives about the nature of reality itself as we buy into the appearance of the world as being separate and materialistic. Our beliefs drive our behaviors, as my dear friend Bruce Lipton says. However, our behaviors in many ways are dysfunctional and they have brought us to the edge of catastrophe. It's about our waking up to unity awareness and taking responsibility for our choices. Otherwise, we continue to play out the illusion of separation and the blame game: "It's their fault," either the immigrants' fault or the government's fault or the multinational organizations. As long as we see ourselves as the 1% and the 99%, we lose the understanding that we're actually together the 100%.

The work that I do is trying to help people heal our collective worldview to support the healing of our damaging behaviors. I'm sharing this wholeworld-view of unified reality to empower a sense of unity on personal, group (including businesses/organizations) and collective levels based on understanding, experiencing and embodying such awareness. I advocate including the experiences and principles of biomimicry. The best organizations, like organisms, are emergent and evolve and can learn by mimicking the natural world. I also bring in cosmomimicry; cosmic principles of consciousness. The third book of the Transformation trilogy after *The Cosmic Hologram* and the second book I'm currently writing, titled *Gaia: Her-Story,* will be called *Many Voices, One Heart* and will I hope incorporate all of this. I'm invited to teach events and give presentations on these concepts in Europe, the US and elsewhere, because there's so much burnout and stress. This leads to the realization that the current approaches to leadership, organizational structures and ways of working are unsustainable and can't go on.

You're involved with Spirited Business; what is that? It's an event that I facilitated in Findhorn, Scotland, in 2019 for its second year. The initiator is a core member of our Unity Community called Jarvis Smith. It's a three-day event where we experientially explore the understanding and embodiment of unity awareness within organizations' evolutionary purpose, to extend that into a sense of planetary stewardship and the sense of expanding the well-being of "me" to include "we" and "all."

My Ph.D. in archaeology was very late on because after I did my master's at Oxford I went into international business for 25 years before deciding to move on. I could feel the beginning of an emergence or a shift of awareness coming through our collective psyche. I wanted to play my part and serve whatever may come through for human consciousness and how we might be able to heal our relationships with ourselves and above all with our planetary home, Gaia. I've been fascinated by ancient and indigenous understanding of realities so I really wanted to bring them into a level of credibility in terms of cosmology, or peoples' worldviews. That's why I researched my archeology Ph.D. at the University of Reading, which has a great reputation for the quality of its archaeological department. My thesis was on the transition from the Mesolithic or Middle Stone Age to the Neolithic or New Stone Age. [*See Steven Taylor.**] People transitioned between being hunter-gatherers and becoming farmers and pastoralists, involving a major shift in how we viewed ourselves and the cosmos with transformations in their perspective of time and space and cultural identities.

Where did the Great Mother fit in in that transition––the Venus figurine fertility figure? Marija Gimbutas wrote a lot about this. We see across the whole of Europe this sense of a primordial mother goddess but now we're finding artifacts that at first were seen to be feminine but actually embody both masculine and feminine attributes––androgynous. When we look at a lot of the artifacts in Malta and elsewhere we see broad-breasted wide-hipped feminine figures but instead of a head they effectively have a penis, melding the masculine and the feminine. It's now seeming that it wasn't so much a shift from feminine to masculine as it was more likely a progressive imbalance of what had previously been in balance. When I've worked with indigenous peoples around the world from Alaska to Australasia, I find that balance of feminine and masculine. It looks as though that balance and associated well-being changed, as the late Neolithic gave way to the Bronze Age when we start seeing more weapons, variations in status, and primarily masculine burials with status goods and weapons, rather than communal ones. The incidence of conflicts also seems to become more frequent and on larger scales as the masculine came to dominance.

Indigenous peoples have given prophecy about the future. Are you finding that playing out in current events? Indigenous peoples have been the wisdom-keepers of our fundamental relationship with our planetary home, Gaia, while we have essentially separated ourselves from her. The South American concept of Pachamama indicates their relationship with her and indeed the whole cosmos, as a family, a web of life, with father Sun and mother Earth, and similarly with Native American traditions that feature grandfather Sun and grandmother Moon. Indigenous people are incredibly sad that we've taken so long to begin to wake up, listen and realize that we need to come together. Some of their prophecies talk about this as the time for all the tribes to do this. Many of them say the choice is ours; we are the ones we've been waiting for. It's time for us to listen and finally to wake up to heal our relationship with Gaia and each other instead of continuing to play out the illusion of separation.

When Deepak Chopra spoke at the IONS conference in 2019 he said the most popular cosmology theory today is the Multi-Universe, but you view the universe as a hologram where the information is stored in places like the two-dimensional edges of black holes. These theories are not irreconcilable. The multiverse is highly speculative but the premise is that the infinite and eternal cosmos has finite universes within it like bubbles or droplets within a great ocean. It may be that there are indeed other universes that, like ours, are formed by cosmic mind in the form of digitized and crucially meaningful information of 1s and 0s, which is the simplest of universal alphabets. From those ones and zeroes all that we call reality is co-created.

The laws of physics in this understanding are in-formational algorithms; instructions to enable our universe to exist and evolve. The evidence points to information holographically manifested as the basic stuff of reality, so that the appearance of three dimensions is actually a holographic projection of cosmic intelligence. The cosmic hologram model fits naturally with a multiverse of possible universes within an infinite cosmos and it shows how cosmic mind "thinks" our universe into being as a great thought. It exists to evolve from simplicity to complexity, a universe knowing itself by experiencing and evolving from its initial simplicity to ever-greater complexity, amazing diversity and individuated awareness. Now 13.8 billion years after it began, we're having this conversation. [*See a paragraph about this emerging framework and the supporting evidence with references on the book webpage.*[3]]

You could say that God is a mathematician? Yes, it does seem that math is a universal language and that the innately relational and patterned nature of reality supports the notion that God's mathematical specialty is number and geometry. *Why is the information located on a two-dimensional edge of a black hole?* Astronomers were looking at black holes and trying to figure out what happened when a massive star collapsed within an event horizon where the gravity is so strong that even light can't escape, so it looks black. Does that information disappear from space-time, in which case all our laws of physics are out the window? It turns out that all that information, instead of being lost, is held on the spherical surface of the black hole pixelated at the minute Planck scale.[4] Each tiny triangular Planck scale "tile" on the surface area of the event horizon holds one digitized bit of information. As the Planck scale is so tiny, there's an absolutely huge amount of information embedded even at the level of subatomic particles.

Just as all the information that goes into the three-dimensional appearance of a hologram is held on a two-dimensional surface, when astronomers saw the same thing happening on the two-dimensional surface of a black hole, they put the two together.

And then they expanded this understanding about black holes to the whole universe as the holographic principle. The 3D appearance of our universe actually is a holographic projection from the boundary of the appearance of space. The answer to your question of why this the case, may be because it may be the simplest way to make a universe.

You report that the universe is a finite, flat toroid which implies that it will end or maybe go back to that little pea-sized shape of the Big Breath. It's flat and finite because we now have good cosmological evidence to support this, but its possible toroidal shape, whilst making sense, is not yet proven. The expansion of space, though, is accelerating which shows that it's almost impossible for it to go back into that tiny little pea. From the holographic perspective that contraction also shouldn't happen because space needs to continue to expand and time to flow to enable our Universe to evolve. The geometric flatness of space, which has been measured at cosmological scales, actually is very important because it enables the equivalence of energy and matter described by Einstein's famous equation $E=mc^2$.

The latest discoveries suggest that tens of billions of years from now our Universe will have its finite end, similar to when we see a bubble grow in air and then it bursts and dissipates. Perhaps at the end of its life our Universe dissipates all of its experience and awareness back into the infinite mind of God? The mechanism for birthing new universes, however, may not be at the end of a universe's life but as part of its "fertility" period when black holes are formed. Young universes may arise from the center of supermassive black holes, such as those that seem to be at the center of most if not all galaxies.

I'm interested in how this explains paranormal experiences like precognition or retrocognition. You write in your book that experiences of non-local awareness are capable of transcending space-time. Within space-time, no signal can go faster than the speed of light, which is crucial for the unidirectional flow of time and universal causality. But our Universe also exists and evolves as a non-locally coherent and unified entity. Think of the Universe as a balloon. Its surface is connected but within the balloon as it's blown up is an experience of perceived separation and causality.

For many years quantum physicists thought that non-locality only existed at the quantum scale but then they started to do experiments and realized that you can non-locally entangle at much greater than quantum scale. They entangled organic molecules and small diamonds. In 2017 researchers at MIT in Massachusetts were able to non-locally entangle photons of light in the laboratory with photons of starlight 600 light-years away. That meant they act as one and aren't limited by the space-time signaling speed of light. In August 2018 they entangled photons of light in the laboratory with starlight and also light from quasars. The farthest was 12.2 billion light-years away, which meant that the light left that quasar 12.2 billion years ago! Entanglement at such vast cosmological scales means that cosmic consciousness is aware of itself in its totality as a Universe and yet it's learning, experiencing and exploring through differentiation and causality within space-time.

The non-local aspect explains our supernormal abilities of telepathy and remote viewing and to some degree precognition. *How does remote healing work?* The future seems to be a bow wave of possibility that comes into the present. My sense is retroactive causation is only possible if it doesn't violate causality within space-time. To use the balloon analogy, inside it time flows and space expands like blowing up the balloon. On the surface we have access to non-locality. What we can't do is to change the past on a causal basis and therefore alter the present because that would violate causality. Within the holographic projection that is space-time there is a signaling speed limit, hence causality, so you and I can feel this differentiation. I can reflect in my garden and the Universe as a whole while at the most foundational level being part of it all. It's unity expressed as diversity, the way in which cosmic mind creates the appearance of our Universe.

The multidimensional realms are nonphysical, so for example it's difficult to communicate with someone who died and is in a realm with very high vibrational level, like trying to communicate with somebody at the bottom of the sea. If we do wake up and remember the oneness of all and our part as microcosmic co-creators of our Universe's reality, we can be more open to such multidimensional communications.

Our thoughts can change matter, as when we do healing, change cell growth, intend for yeast to release more oxygen, etc. Of course, because everything is mind and consciousness, articulated as meaningful in-formation and expressed through thoughts and emotions. We're like babies in terms of our realization of this and yet in ancient wisdom teachings there's a lot of guidance that basically says unless someone is mature enough to be able to deal with these abilities, there will be a natural limitation on their use. These are incredible powers that would have been attributed to gods, but that require wisdom rather than knowledge to utilize.

What organizations are doing cutting-edge research? There are a number and with whom I'm delighted to be involved with. I'm an advisor to the Galileo Commission, part of the Scientific and Medical Network, whose recent Report is entitled "Science Beyond a Material WorldView." I'm also a scientific advisor to the Global Coherence Initiative, supported by HeartMath [*see Rollin McCraty**], doing research into consciousness and heart-based coherence on personal and collective scale and how this may resonate globally with Gaia's informational field. I'm a member of the Evolutionary Leaders circle whose aim is to be in collective service to conscious evolution. And in 2017 I co-founded WholeWorld-View that has grown to a Unity community, which I co-convene, of around 900 global changemakers who support the aim of understanding, experiencing and embodying unity awareness and celebrating it through diversity in service to conscious evolution.[5]

A lot of folks are embodying unity awareness as foundational to their evolutionary purpose. It's become a sort of a byword for the integration of science and spirituality and the perspective of unified reality. *To define consciousness, from reading your latest book, you equate it with information?* Essentially consciousness articulates itself as meaningful information––in-formation. In English, we have an alphabet of 26 letters, which of themselves have no meaning. Our Universe has a universal alphabet of just two letters; ones and zeros. Just as our technologies convert our human languages into long streams of ones and zeros and then recreate the words, our Universe does it from cosmic intelligence. Consciousness co-creates the appearance of our Universe from the universal alphabet of ones and zeros. Just like our human language of words and sentences, they're brought together as meaningful patterns and relationships that literally co-create and in-form the appearance of space and time, energy and matter.

Materialists are reluctant to tie quantum non-locality into something like telepathy. They say, "We don't know what that connection is between entangled particles, so you can't say that it leads to telepathy." How would you respond to that argument? I don't think they follow where the evidence leads. The meta-analyses that have been done and all the work that IONS, HeartMath and many other groups and research organizations have carried out show that supernormal phenomena are very well evidenced. However, it's impossible to account for them in the mainstream paradigm of a materialistic and separatist universe. We know though that our Universe is non-locally connected because quantum physics doesn't work if it's not. The evidence for a non-local connected Universe within which there is the appearance of separation, for all sorts of good reasons––for consciousness to know itself and experience itself and differentiate itself––is very powerful. Supernormal phenomena are then a natural attribute of our consciousness as aspects, as microcosmic co-creators, of a wholly entangled, in-formed and holographically manifested Universe.

Let's go through your books to see how they've evolved. The first one was The Wave: A Life Changing Journey into the Heart and Mind of the Cosmos *in 2005.* I had no intention to write any books but publisher John Hunt approached me and kindly said, "I love what you're writing." I'd written a couple of very short articles by that point, so we met, and *The Wave* came out of that. What I was really enjoying at the time was teaching, I led sacred pilgrimages around the world, working with the planetary grid and teaching experiential workshops in sacred geometry. *The Wave* responded to what was flowing through me that I was enjoying at the time. *How did you discover where the grid was and the power spots where you tried to make things more harmonious? Did Thoth help you?* Thoth and many other beings helped. You have to realize this has been a journey of six decades. The understanding came through a variety of direct guidance, research, reading books, and linking with and learning from other folks who were doing it and wisdom teachings of many traditions. It came from a great deal of multidimensional communication, including directly with Gaia as a living being.

Gaia doesn't look like it's doing very well, with global warming, increasing inequality; do you have hope? I wrote *Hope: Healing our People & Earth* in 2010 just after Obama became president of the US and at the time of the Arab Spring. When I first started to write it, I was feeling quite hopeless, but then, as I started to write it, it was like seeds of hope started to rise within me. That hope it seems to me is offered through action, healing, re-membering who we really are and remembering the oneness of all creation. We're microcosmic co-creators of a living planet evolving within a living Universe. We're in the middle of a collective healing crisis at the moment; as for an individual in a healing crisis, we can either recover or die. We need to remember our relationship with our mother Earth who is our primary mother.

Do you look at the autocrats as bringing the shadow of racism, sexism, classism, and xenophobia up to the surface? Do you see them as part of the healing? I do, in the sense that I do see them as proxies, reflecting back to us those aspects of our collective psyche that is not healed and yet needs to be acknowledged, for the possibility of healing to occur. I wrote in *Hope* about the unhealed wound of the abandonment pattern that plays out through the perception of the American Dream. The pattern is carried by many immigrants fleeing conflicts, poverty and other circumstances of being "left behind."

Rather than healing from within, the culture, it seems to me, is continuing to try to feed itself from without. You can never have enough stuff if you try and do that, plus the unhealed wounds of slavery and genocide of Native Americans are also still playing out in the collective American psyche. I wasn't surprised at all when Donald Trump was elected, because I could see the people who'd been left behind through globalization. The US in many ways lacks the social support we have in the UK and other places. I believe that it was part of a potentiality for healing that this happened. If Hillary Clinton had been president, I don't think that the existing complacency would have been broken.

Let's talk about The 8th Chakra. The eighth chakra is midway between the heart and the throat. In my 20 years of experience with it, it's the higher heart, the universal heart, a level of vibrational awareness, the bridge between the personal sense of self and transpersonal sense of self. It's a bridge between the seven chakras of our persona, our personal sense of self and ultimate unity awareness. The ninth chakra is beneath our feet, the tenth chakra is above our heads, while the eleventh and twelfth are ever higher, more expansive levels of awareness and relationship with ultimate oneness. The ninth chakra is sometimes called the Earth Star by planetary healers, a wonderful resonant communication channel with the intelligence and consciousness of Gaia. The tenth connects us with our "soular" system, so as a soular astrologer, I communicate at that higher level of the archetypes and the logoi of our solar matrix of consciousness. The eleventh chakra is a galactic level of awareness, the twelfth is a universal level of awareness. When all twelve are activated, unity awareness of unified reality is fully embodied, as Christ-like or Buddha-like perception. The book tells how we can expand into each of those levels of unity awareness, which I believe is our evolutionary potential.

The 13th Step is a companion piece to *The 8th Chakra* because it tells the true story of 13 global journeys taken to activate the planetary grid to serve the potential for a collective shift of awareness. It's that inner remembering of wholeness on personal and collective levels, of which the eighth chakra, the universal heart, is the bridge into that healing and wholeness. The journeys began in 1998 when I got a download for 12 journeys to specific locations around the planet. When I plotted them on a globe, I found that they were each within the 12-faceted dodecahedron planetary grid that Plato had talked about 2,500 years ago. They began in Egypt and then to South Africa, China, Alaska, Peru, Australia, New Zealand, Antarctica, Easter Island, Hawaii, Reunion Island, and Avebury (an ancient site near where I live in England). As I was invited to write the book by Hay House, I had a guidance of a thirteenth journey, which completed in Jerusalem to invite the Divine Feminine of the Shekinah essence back into her activation to guide and nurture such a shift of consciousness.

Did you feel that Gaia shifted as part of your connective process? It wasn't just myself, because I traveled with nearly 70 people over those journeys and it was our feeling that, yes, a shift occurred. In the decade since the book's been out, I've had so many people come to me who are doing their own healing work around the planet and saying either "your stories inspired me" or "they set the scene, they began to anchor this connection." It's very much a wave of change as we're finding the way people are relating to Gaia. It's the story of so many people over so many years who undertake pilgrimage and serve as healers of the traumas that we, in our misunderstandings and our fragmented perspectives, have imprinted on her awareness.

What is the topic of your next book? It's the story of Gaia herself and her heritage going back 13.8 billion years to the beginning of our Universe. It's the most amazing story. This wholeworld-view, of which the cosmic hologram is the framework, is really the science of consciousness. It's the science of love because for me love is foundational and connects everything. The incredible love of a cosmos that births a Universe such as ours, embodying an innate evolutionary impulse to experience and explore. It's not a sentimental perspective but everything in existence has worth and value. *Is Thoth helping you write the book?* A number of amazing beings are helping me so it's very much a team effort. This is the voice of Gaia that requires me to do deep listening to perceive her voice and her guidance about her story.

Does it make you hopeful that young climate activists like Greta Thunberg are raising this issue? I'm a realistic optimist but it shouldn't have taken a 16-year-old to raise this issue. I am ashamed of my generation for being silent, but I'm not going to participate in the blame game. My work is about linking up and lifting up to welcome everyone who wants to be part of regenerative and restorative action. Our beloved Gaia is after all our first mother.

My book *CosMos: A Co-creator's Guide to the Whole World* with Ervin Laszlo was one of the first times I set out this idea of reality being innately and meaningfully informed and holographically manifested. We discussed future trends from a co-creative evolutionary perspective. It's down to the choices we make. By buying into a myth of separation we behave accordingly, so everything we do comes from that perspective, whether it's unnecessary scarcity or conflicts. Using biomimicry we can learn from what Mother Nature does if we're willing to get off our arrogant high horses and with our complicated technologies. Gaia doesn't waste anything or pollute anything. Everything is part of the intermeshed web of life; by hearing and listening to her we can learn so much and perhaps to not only survive but to thrive and consciously evolve as a species.

What do you do for fun in the face of this deep thought? I'm a great film lover; I love gardening and being in Nature. I love culture and ancient wisdom. My entire life is about curiosity and the wonder of the Universe fills me with joy. And I love music and dancing.

Books

The Wave: A Life Changing Journey into the Heart and Mind of the Cosmos, 2005

The 8th Chakra: What it is and How it can Transform Your Life, 2007

CosMos: A Co-creator's Guide to the Whole World, with Ervin Laszlo, 2008

The 13th Step: A Global Journey in Search of Our Cosmic Destiny, 2009

Hope: Healing Our People & Earth, 2011

The Cosmic Hologram: In-formation at the Center of Creation, 2017

Fiction: *Heritage of the Dragon trilogy*, 2014 (free ebooks)

Endnotes

1 https://cafeastrology.com/articles/aspectsinastrology.html

2 Monica Gagliano. *Thus Spoke the Plant: A Remarkable Journey of Groundbreaking Scientific Discoveries & Personal Encounters With Plants*. North Atlantic Books, 2018.

3 https://visionaryscientists.home.blog/2019/12/16/jude-currivans-theory-of-reality/

4 https://van.physics.illinois.edu/qa/listing.php?id=1277

5 https://www.wholeworld-view.org

Brian Josephson, Ph.D.

A Unified System

Photo by Professor Josephson

Questions to Ponder

Dr. Josephson's research interests seem different but are unified by his search for a unified system, coordination, or design. Explain.

Why did he conclude, "My work on the mind is more important than my Nobel Prize-winning research?"

How does he define Mind as different from the brain in a way that goes beyond quantum mechanics?

The author faults contemporary scientists for doing science by consensus. What four taboo topics does he think deserve study?

You are a Capricorn. Do you identify with being a mountain goat in any way? I haven't studied astrology. *You have a history of speaking up when something isn't truthful and that takes a lot of courage and forthrightness. I think of Capricorns as being like that.* I think it's just my type of personality. I was always curious. My parents were quite intellectual and I had intellectual discussions with them so I got used to intellectual argument. I got used to thinking about what people say, so I would decide whether I agree with something or not agree. I didn't think of inhibiting what I was saying as I would just speak out as one would in an ordinary conversation.

Your father taught French, but he was interested in math. Did he trigger your interest in math and physics? In part, but I think I was interested in mathematics anyway. I found it a fascinating subject to be able to prove things in geometry and so on. In physics my teachers were very helpful and gave me books to study and assisted me in various ways. *What was your birth order in your family?* I was an only child. *What about your Myers-Briggs personality type?* I think I was especially introverted but I've become much more extroverted, partly due to meditation.

Your interest in parapsychology started when you met your Trinity College colleague George Owen, who was interested in poltergeists and PK. Why did you think he wasn't crazy? I didn't think a colleague of mine would have taken it seriously if there wasn't some kind of evidence. *PK is pretty specific because someone bent your desk key. I've done spoon bending; it kind of melted and curled in my hands.* I've done that to some extent. It was a relaxed session and it didn't seem like you were doing it by force. *If you were hazarding a wild guess, why would you say we can do PK or how your dissertation advisor's mother could know how her soldier son was doing in WWII?* I think it's a basic principle that our thoughts connect with the physical world and maybe a question of developing the skill. That's a general principle and I suppose it may to some extent be up to how much you work at it. Unless you think it's something that may work, you won't develop the skill.

An overview of your research interests seems very disparate:

Quantum tunneling and the Josephson effect
The observation or measurement effect in quantum mechanics
Semiotics/Language construction
Intelligence in nature—semiotic scaffolding
Cymatics (how water responds with patterns to sound waves)
Water memory (explains homeopathy)
Music and mind
Parapsychology and Mind

You wrote, "Mind processes are relevant at levels other than familiar ones such as brains and are involved in contexts such as quantum observation."

A word that repeats in your thinking is "systems." Is it fair to say that you're looking at intelligent patterns, a kind of modern look for Platonic forms that precede matter? What Pierre Teilhard de Chardin calls the noosphere, Rupert Sheldrake calls morphic fields, or David Bohm's Implicate Order or Wheeler's "fabrication of form"? Is this the unifying theme? At the beginning of the 70s, my interest shifted from regular physics (superconductivity, etc.) to the brain. Also, through my friend I got to know about parapsychology. I started to be interested in the mind and brain so my interests, such as parapsychology, came from that. I had a colleague who I discussed music with and how that fitted with mind in a meaningful way.

Let's define brain as opposed to mind. Brain is the little fatty, liquidy thing we have in our skull, but mind is consciousness? I wouldn't say consciousness. From my perspective, the brain is a physical thing connected to mind because mind does the thinking, perceiving, and so on. It's a bit like what a computer does, but perhaps more intelligent than a computer although there's something basically physical about it. It's the opposite side to matter. They're both parts of reality but mind is something different, which physicists hardly study at all, but I think is very important. *This goes back to Platonic notion of forms with the ideal form that exists before matter patterns itself on that idea.* I think it's beyond what we really understand. David Bohm talked about infinite levels of meaning. Physics doesn't tell us much, but by reasoning about how the mind works and what it can do, we can do better. I'm hoping to make it more precise and more like physics, even if it's not like the physics we have at the moment.

Physics relies on mathematics. Are you suggesting there might be a mathematics for what Jung calls the collective unconscious, this big intelligence system? Yes, but mathematics is a bit of a mystery once you get to abstract mathematics. Language is equally mysterious. I think there is some kind of orderly process going on which leads to these things. One idea floating about is the idea of Coordination Dynamics, the idea that processes work together to produce phenomena. I think this works step by step, producing more and more powerful effects. Once we've got the idea, we can try and play with it to work out the details--at least, that is my hope.

Is that what religions would call God or Divine Intelligence? Is your concept of mind analogous to God? Not necessarily. We don't know that. We have to look and see where it leads. It's not clear what will convince people. *Mind is a system, an intelligence that underlies evolution? Does that describe this big concept of mind, not my individual brain?* Yes, I would agree with that because I think there is something to intelligent design. Evolution isn't as random as people say. The basic form may be predetermined. Just as we have plans and then work out details, I think nature/ God has evolved a plan.

You point out that mind processes are relevant at levels other than familiar ones such as brains and are involved in contexts such as quantum observation. What's the implication of quantum observation influencing matter? The situation is not very well defined. There are lots of things that aren't very definite but they become definite when you do an observation. It shapes the system, which is in essence ill-defined. *What are the implications? It's like that old saying about if a tree falls in the forest and no one hears it, is there really a sound of a tree falling? Does it mean that existence depends on our interaction with matter?* Probably not. Karen Barad (author of *Meeting the Universe Halfway: Quantum Physics and the Entanglement of Matter and Meaning*) has talked about this sort of thing. There is some part of nature personal to us but part of it is more objective in character. There may be something that imposes a fairly stable structure upon things that makes them more objective, less affected by what we do but maybe not totally resistant.

The fact that the researcher influences the outcome of the experiment means that the Western paradigm of objective scientific research is false. Do you think that scientists fully understand that? Normally we don't think about quantum effects––the world is just there. So, it's just what we look at in experiments. I suppose quantum physicists are aware of that. There's the question of how much we influence an experiment without having that specific intention–– that is something that parapsychologists think about. Ordinary scientists, quantum scientists, and parapsychologists all have different views. Some things are more sensitive than others: That is the moral. That understanding should be included in the ultimate physics that we are working towards.

You've criticized science for being science by consensus. I think there are prejudices against psi, while a lot of people are somewhat open-ended, saying, "It may be true but I haven't seen anything that convinces me." Then there are people who have seen evidence and are convinced. *When you read the skeptics, what are their common arguments against psi phenomena? You can't just say it's weird because quantum mechanics is weird.* Yes, it's just an alternate faith really. I don't think they have good arguments against it. It's just a strongly held belief. *Have you read any argument that seemed to be logical against psi phenomena?* There can't be a valid argument against something that is real.

Is it typical that when there is a paradigm change, the old guard has to have one funeral at a time before it's accepted? Peers have been afraid of your interest in parapsychology and other people get a lot of flak, even from something like Wikipedia. Wikipedia is the last place to look for objectivity in this kind of situation, as it has been taken over by the Guerrilla Skeptics (volunteers who aim to increase skeptical content) and the like. It will take some time to get sorted out, like the quantum theory and continental drift. The latter was rejected at first, but gradually the people working in the field came to realize it was valid. In the case of physics, perhaps the most distinguished person is John Wheeler who presented the idea of observers constructing reality, but people tended to ignore that side of things, similar to David Bohm's ideas. I don't accept quantum physics as fundamental. One way to put it is to say that vibrations are important; it is a part of the whole scheme because if anything is coupled, it's through vibrations connecting together, at least that's one idea. *Is that string theory?* There may well be a connection.

Time isn't what it seems as evidenced in Dean Radin's experiments where he showed people's slides and they reacted to the disturbing slides before the computer even selected them and remote viewers can draw the target before it had been selected. Perhaps this is all one big system of things working together and perhaps things are therefore somewhat determined in advance. *The remote viewers would accurately draw a target before it had even been selected, so there's nothing deterministic about that because it doesn't matter which target is selected.* Maybe it's in some way determined what the random event is going to be and people are tuning into that. So once you have this strange kind of organization, I'd expect that anomalous things can happen. *Are you saying that since they focus on the target that generates the target being selected?* Yes, maybe something like that. That's all a bit speculative so you may say it's not clear yet but we are working on it.

You got the Nobel Prize for the tunneling effect that led to faster computers. Did that change your life, to win the Nobel Prize? I think the Nobel Prize helped a lot because it meant people couldn't easily get rid of me. Before that I had to be a lot more cautious. The powers that be couldn't really dismiss me or insist that I work on conventional things but they discouraged students from working with me. Usually people have a lot of students enabling them to develop their ideas much faster, so it's taken decades to get where I am now.

You started the Mind-Matter Unification Project at Cambridge; what is its goal? To develop these ideas but I haven't got many people working on it. There's a certain hostility in the department. A student was told she could have a grant from the department if she worked with anybody other than me. I did have a student who wasn't put off by the pressure, but he did something more conventional so that didn't help the project. Now that we've got e-mail, there are a few people I collaborate with in that way.

Some say quantum entanglement and non-locality explain why we can do ESP or healing from a distance or clairvoyance. But I think physicists like you say, "No, we can't make that conclusion." Yes, that is the case. I think we have to introduce mind into the picture to get any psi phenomena and that has something to do with coordination. That is the next level of understanding of nature to see how all that works. One case I've considered is a computer simulation of how language works by Terry Winograd. You see in that case a lot of things working nicely together to produce a skill such as understanding. I think that is the general process, a general mechanism that can do things and produce particular kinds of capabilities. In the case of matter, this mechanism will produce the regularities that we associate with ordinary physics. *It's some kind of connection that involves communication and maybe thought having an impact on matter?* Yes, it's very much like an ordinary skill. It's like walking on a tightrope. The phenomenon will not occur unless you make various processes work together. We have to work out the details and models.

Have you had any inspiration about mathematical models for Mind with a capital M? I had a student working on that once. He took a very simple case of learning balance in one dimension. He had a computer program that learns how to balance simply by trying various combinations of agencies and seeing how to combine the parts, firstly to reach a vertical and then learning to balance. So, you can produce computer models. *Was this balance done on the screen like a drawing or was actual three-dimensional balancing going on?* It was purely a computer model in two-dimensions. *The point is there was a system involved in learning to balance and that could be modeled on the computer mathematically?* Yes, this was an idea of Norwegian mathematician Nils Baas called hyperstructures. The computer model that tested the theory wasn't just a set of abstract equations; it was a proper computer simulation. But despite its initial success, continuation of the project was blocked by the department.

You were quoted as saying, "My work on the mind is more important than my Nobel Prize-winning research." What particular aspect of the brain/mind will make the most impact? The work I am doing now, as in my paper on the physics of mind and thought, putting all sorts of ideas together and hopefully taking physics beyond quantum mechanics. There are two main approaches to understanding nature, that of the physicist based on mathematics and that of the biologist focused on function, and we need to bring these two cultures together.

Biological scaffolding—Is that the kind of concept you are interested in, which has implications for physics as well as biology? Yes, that's the idea. It's the biologists or biosemioticians who produced that idea. I don't know if it's anything other than an explanatory concept. The people working on coordination dynamics have been working on mathematical models, but rather simple ones. They already have equations for how coordination occurs. *Could you give us an example of coordination? Is that like a beehive where the whole system works together? Every worker bee or queen or drone knows what to do?* In biology, you have lots of things happen which must depend on individual processes working together. For example, walking requires balance and stepping. Speaking involves a whole number of things working together. That is a principle that goes beyond what people have in physics who have a very simple example of coordination. In biology, it's a more complicated business. *It requires thinking in terms of a system, many parts working together in a coordinated, intelligent way?* That's right, yes.

What does scaffolding mean to biologists? That is the thing that does the coordinating. Conventional theory is you have the connection between genotype and phenotype, genes and phenomena, but there is a question of how it actually happens. The biosemioticians say we need to look in more detail. We can't just say we have a code that does certain things because it involves an additional organization that you have to put in before it happens. The organization may actually be what we talk about in epigenetics. The activity may involve organization maybe even in a single cell that gets passed on from one organism to another. It's a lot more complicated and involves new ideas. Scaffolding is something that helps you do the right thing; you do things in an organized way because there is something that does it. *It involves some kind of intelligence and memory?* Yes, if you like.

What has helped you keep centered in the face of opposition to your interest in these unusual topics? You started transcendental meditation in the 70s. Yes, I was at Cornell University in New York and a Beatles follower when someone told me about a meditation group, so I thought, "I'll go along." The head of the group said, "You know, it may change your life." So that's what got me into it. *How has it changed your life?* I have more awareness of the world and clarity in my mind and meditation is one factor in how I've managed opposition to my research. Later I was invited to go to a place in India with a different kind of meditation, that of the Brahma Kumaris, where I felt the intensity of silence in the ashram and thought, "This seems pretty good, I'll try it." I am more alert during meditation and need less sleep so I have continued it ever since. I also find music helps me go to a deeper state of consciousness. I choose a piece of music that I think is appropriate to my state of consciousness and then it's a question of awareness in the presence of music. I changed to Raja Yoga in the 80s.

Have you found that meditation is a source of information for you? I think I always had access to a certain state of consciousness. When I learned about *Samyama,* which is a combination of three states of mind, I realized that it was a process I had been doing all my life to get insights. I suppose meditation helped. *What are those three states? Dhāraṇā, Dhyāna,* and *Samadhi,* which are concentration, thinking and transcending thought. You combine them all in some way. If you concentrate on something, then you start thinking about it and then you transcend thought to let inspiration come. *When is the optimal time to receive that downloading?* To some extent walking, but also when I'm semi-asleep in bed. I find things become clearer during the night and then I can write it down in the morning.

Have you had children along the way? Yes, I've got one daughter and one grandson. *Is she interested in science?* She wasn't keen on physics; she is a veterinary surgeon now.

Could you explain your list of significant topics that scientists dismiss? One we've talked about is parapsychology. Two, I also mentioned intelligent design, which is confused with creationism so is dismissed, but it's really a different kind of faith to assert that evolution *doesn't* involve any element of intelligent design. Three is the memory of water. As it happened, I met Jacques Benveniste at a conference prior to the trouble he had with the *Nature* journal and we interacted on the subject. It's been taken up more recently by the Nobel Laureate Luc Montagnier. Basically sound can structure water in some way and give it specific properties. That might be clear evidence. I don't know how many people have replicated cymatics, which is viewing the effect of sound on water that produces patterns. That's clear evidence that something interesting can happen with water. *I read that Dr. Masaru Emoto's studies of water crystals in Japan have been replicated.* I've not heard of replications myself, but in any case paranormal phenomena are easily disturbed and may be influenced by the experimenter. A famous physicist, Wolfgang Pauli, had a reputation for disrupting any experiment he was near.

Homeopathy has been used widely in Europe since the 19th century with success. However, many people are very strongly against it, generally on the basis of faulty arguments, such as saying that the water is so dilute so it can't do anything. They ignore the fact that water itself might have a structure. *Since our bodies are about 60% water, what are the implications for our health that water perhaps has memory and structure?* Structure of water may be very important biologically. It's surprising it has not been taken seriously because structure is important in chemistry so why can't structure of water be relevant to biological function? *It's another one of those "science by consensus" problems.*

You've also been interested in cold fusion. Awareness is growing that we are destroying the planet with fossil fuels. If we had cold fusion nuclear reactions at room temperature that might be an alternative to fossil fuels. Cold fusion is a real phenomenon, demonstrated by many experiments but it's not understood at all why it happens. [*See Garret Moddel.**] People get it to work by trial and error, but not reproducibly. There is no particular reason why it should be reproducible, which is often the case with materials. The process, when it occurs, may be connected in some way with coordinated activity. There are theories but it's all a bit mysterious and really there should be more support for it, so that there can be more research. Perhaps what happens is that when people start to get good results they get quiet about it because they are hoping to make money but I don't think people are making very large amounts of energy from cold fusion.

Are you hopeful that we can save the planet or do you think we've gone too far and irrevocably destroyed our habitat? It should be possible but the problem that is being talked about a lot is that capitalism is messing things up and we need to think of an alternative. The trouble is politicians are influenced by the forces of money and it's difficult for them to do the right thing and also they tend to think in the short-term rather than the long-term. Knowing the power of words, people think up phrases to influence people like "strong and stable," which didn't work very well in the UK. Psychologists are at work, trying to influence people.

Selected Paper

"The Physics of Mind and Thought." https://doi.org/10.1007/s41470-019-00049-w

Menas Kafatos, Ph.D.

Consciousness is the White Elephant

Photo by Susan Yang

Questions to Ponder

How does Dr. Kafatos define consciousness and how is he influenced by ancient philosophies East and West?

What's one of the most momentous recent scientific discoveries?

Our social problems are caused by linear thinking. Explain.

I was born in Crete in Greece on March 25th, 1945. I'm the third child of three sons. *Do you identify with being an Aries?* I would say yes, in terms of leadership and looking forward. But of course, I'm not just Aries since like everyone, I carry other parts of the zodiac, an E/INFJ, similar to the group profile.

What was your family's attitude toward religion and spirituality? My mother was very religious Greek Orthodox and my father, a free thinking individual, in many ways was as spiritual. I distinguish the two because religion is more formal while spirituality is more universal but the two are not opposite to each other. As a young boy, I respected and participated in functions of the church, I found them fascinating and fulfilling.

How much of your childhood did you spend in Crete and how did you get to the US? I spent all my childhood in Crete. Before I came to the US, I had travelled only to Athens. Then at 18 years old, I came to the US and then I stayed. I studied at Cornell and then at MIT. *How did you get such a good education that you were accepted to Cornell and MIT?* I was in a private school. I owe a lot of it to my father. He came to the US at the beginning of the 20th century and became an American citizen but eventually went back to Greece. All three boys grew up learning English from our father and French was my second language. My middle brother, Fotis, was a well-known biologist. He went to Cornell and studied biology and then went to Harvard and became the youngest professor there in the 60s. Early on, I started as an artist because I was born with a gift of drawing and painting so it was easy for me. The plans early on were for me to go to Paris and study fine arts but at age 14 I told my father, "This is too easy for me. I can do this with my eyes closed. I want to do something challenging. I want to study science." He said, "In that case, you have to go to America and you have my blessings." I got a scholarship to Cornell and ended up at MIT for my Ph.D.

What was the biggest culture shock when you went to Cornell at age 18? It wasn't much of a culture shock because when you're 18 you are very open. My biggest shock was when President Kennedy was assassinated a few months after I arrived in 1963 and then in 1968 Martin Luther King, Jr. and Bobby Kennedy were shot. I've never gotten used to the gun violence in America. *And now the epidemic of school shootings, etc. is a major problem.* It is indeed a problem that we all, independently of where we stand, need to address.

When you were getting your Ph.D. in physics at MIT, what would you say the dogma was in terms of parapsychology and psi phenomenon?

Did they even talk about it? Not much in those days. My mentor was a top physicist, Philip Morrison. We were interested in questions about cosmology, supernovae and the quantum––how do you bring the two together? Then I started looking at philosophy as the foundation of science, coming from the Greek tradition, not the other way around. This led me to metaphysics and where I presently am in how I view the universe and myself.

Most of the grad students and profs at MIT were men? Yes, but that is not true anymore. When I went to George Mason University and became a professor in the Physics Department, the majority of the physicists were women, which was indeed good. *It's still true that most of the engineering, physics, and computer science graduate students are men.* Now neuroscience is 50/50, the so-called "soft-sciences," the life sciences, tend to be at least 50/50, if not dominated by women.

Do you think that there are any gender differences in approaching the study of consciousness and metaphysics? Are women more inclined to be open to it, or not because they have to prove that they are part of the fraternity? I would say that as far as scientific acuity, women are at least as good as men. I say at least as good because my experience has been that women are a little bit smarter than men but that's a different story. *Maybe because they know they have to try harder and have less ego that blinds them?* I think that may be part of the reason, but I'm really concerned now because I see men in several countries are sometimes falling behind women. Women students tend to get better scores and tend to stick it out. So, eventually, we may end up having most departments with female professors and not many male professors. We need a balance. *Around the world, there are more women in university, except in Africa.*

The women's movement said to women, "You can have it all," but the subtext was you've got to be good-looking, have an excellent academic record, a prestigious profession, and so on. They knew that they had to work hard. I can empathize with that as being from a foreign culture, I had to work very hard. I tell young people, don't talk about your brain, talk about hard work if you want to advance. Women know that from childbirth all the way through life. Men are falling behind because perhaps society gives too much to young men so they become a little lazy.

What research do you do in South Korea? I collaborate with Korean scientists in the area of climate change and natural hazards. They pay much more attention than we are in the US in terms of government policies, relative funding, and concern about the environment. Of course, California is a different situation. We are happy to live there. I give lectures and work with groups for youth to promote consciousness to understand where we are in the universe. We need to wake up to our true nature instead of thinking ourselves so small.

Do you find that youth are more understanding of that concept? They are certainly activists around the world to stop climate change, especially led by girls like Greta Thunberg (my next book is about them). I think in South Korea younger people are more open than older people. In many other countries as well. In my country of origin, Greece, people of all ages are concerned and are involved. I consider myself young in mind even though I'm well into middle age. I would say that young people are looking, are seekers--the same in Korea and Greece and many countries. (I travel a lot to Korea and Greece.) People are looking for a new paradigm because they know that the old paradigm is not working. This separation and division does not work and young people are concerned about their future. I have three sons who are concerned about their future too.

What were the most difficult challenges in your life and how were you able to get through them? One of my biggest challenges had to do with my personal, but I think I've become wiser because of the whole process. At the professional level, I could always see beyond the horizon into the future into things that would develop. I was the dean of my college at two different universities for 10 years; it was hard sometimes to convince my fellow professors that there is more than what we teach. In terms of being an innovative dean, I had the backing of the administration at George Mason and then at Chapman University where I am presently. Sometimes it was hard to convince my fellow faculty members that we should be open and can't be too narrow-minded in academia.

You're married now to another professor. Did you learn from your first marriage in a way that helped the second? Yes, my favorite line in *Indiana Jones* is when he says, "You chose wisely." Although I view my entire personal life experience as positive, I can say that I did choose wisely the second time. *We have to go through different experiences to expand and be challenged.* I'm married to a Korean neuroscientist. She is a very active modern person and is a little more practical than I am. Women tend to be more practical, but certainly she is very open-minded.

Do you have a meditation practice? I meditate every morning, mostly using a mantra and following the breath. For example, I use the mantra *Om Namah Shivaya,* which is a universal mantra of conscious awareness, and *Soham* or *Hamsa,* which have meaning in Sanskrit, they mean "That I Am," "I Am That."

People report that once they commit to the spiritual path, synchronicities speed up, just as you write about. Absolutely, yes. What came up in 2018 in Korea talking to young people, I started to sense some information that I knew was not really coming from me. What came out were eight steps, principles, one may say laws that apply to everyday life. The first is to expect the unexpected, which is very much a quantum mechanics principle. Synchronicity works through the unexpected. You look back and understand, "That was not random. There was a purpose behind it." The second step is pay attention to the little things in your life. The third step is we see the results only after actions. The fourth step is to tell your mind and ego to wait! By the time you get to the fifth principle, one step at a time, do not expect big miracles; look for little miracles in your life. When you follow these five steps, in the sixth step, things start moving very fast. Time expands so you can fit in more. We know that there are two types of time: cosmic time and clock time, the subjective time that we live by. You find that your subjective time expands and you can do a lot more and you find these little miracles taking place every step of the way. The seventh and eighth steps have to do with the end of life and what seems to be a common experience, the difficulties we encounter in our lives.

The question is, "Who is guiding us?" The answer, I think, is the universe. *Does that mean a universal intelligence? God? Personal guides?* All of the above. I think there is only unity awareness, only one consciousness in the universe and we are a part of it. I don't talk too much about God as a word because that is more of a personal path and people generally get excited when you talk about their specific god, so I prefer to call it consciousness. It's of course the same idea.

People talk about a new paradigm of non-materialist consciousness, but Greeks were talking about idealism in 500 BCE. Plato talked about the form that's behind the material manifestation so it's a very old paradigm. Yes and I also studied the Vedas, non-dual Vedanta and Shaivism, some of the great Monistic schools of India. It's an ancient paradigm, but we forgot it because of the materialists and now we're rediscovering it because of quantum mechanics. Heraclitus was the first quantum philosopher because his remarks on the nature of reality were very on target. And of course Plato, Socrates and the Ionian philosophers are all part of a rich tradition in the West, followed by the Neoplatonists and mystical Christianity. In the East, we have the non-dualistic schools of Advaita Vedanta and Shaivite systems, which are very sublime systems in the ancient Vedas. Also, I look to the Chinese Taoists and Japanese Buddhists—particularly the Zen Buddhists. The message of all these different traditions is we are more than we think we are. I think quantum mechanics can give us the tools to understand reality scientifically.

Would you say that since Descartes, the West separated materialism and spirituality? Yes. *And then quantum physics is turning it around.* Following some of the Hindu traditions, it turns out that the creator Shakti is actually feminine and Shiva is the male masculine aspect, the underlying reality responsible for shock that creates the universe. So, of course, if you're a pagan all the way, it is women who create life, so we have to put things back into perspective. *Native Americans revere Corn Mother, Buffalo Cow Woman and so on. I think the West is unique in being so patriarchal and male-centered in its view of divinity. Catholics have Mary, but that's about it in the West.* Right, the Catholics and the Eastern Orthodox church pray to Mother Mary to intercede with Jesus. *But being an intercessor is not the same as being the one with the power, as in, "Go ask your mom if dad will give you the car keys."* Speaking about that, my father grew up in a village in Crete where my grandfather was the male patriarchal figure. But my father would joke because at night when the kids were in bed you could tell who the boss was. My grandmother would tell her husband, "What are you doing? Why are you doing this?"

Let's consider an easy question: What is the hard problem of consciousness? It's where does consciousness come from? If you try to build it from the bottom up, it's an impossible problem. So, I take it the other way around from the mystical traditions of both the East and the West, indicating that the science of the future will accept consciousness as primary. Then, it's easy because it's all driven from universal consciousness, an idea, a plan. Look at the Parthenon that uses the Golden Mean, the principles of mathematics. The point of view of the great plan is the easy problem, while trying to create conscious awareness from the bottom up is impossible.

People use "awareness" and "memory"; what are other synonyms for consciousness? In the English language, unfortunately, we only have a couple of words—consciousness, awareness—but in the Sanskrit language there are at least 20 or 30 different names for Shakti and Shiva. *Natandrya* is another name which means absolute freedom of the universe. All these different names represent different aspects of the goddess. Usually, when we say "consciousness" in the West we mean being conscious of something else. I like the term "awareness" because that is the underlying part of self-consciousness as well as object-consciousness or awareness.

Is a single-celled animal aware that this is too hot or cold? Slime mold have intelligence and fungi solve problems as shown in the film Fabulous Fungi. *Do trees have consciousness? Where does it end?* Once you go down that path, even rocks have consciousness. Minerals have consciousness, but not human consciousness. The chemical elements that form a rock are also in our bones, so what makes the difference between a rock and ourselves? In the case of humans, it's the self-awareness. But if you view universal awareness as taking different forms, then rocks are part of the big picture. The big question we have in science is where do you transition from living to non-living? Without DNA, you could not have genetic code. So, is DNA living or non-living? I think the transition is not as sharp as sometimes we think it is. *DNA is alive. It changes form, but it can't exist on its own outside of a cell.* Exactly and human beings cannot exist on their own either. We're part of the big picture. One of the big problems we have with climate change is the Western viewpoint that humans have dominion over mother nature which is not a good concept to have. I think the Native Americans have a way better attitude towards nature than we do. *Right. They partner with it with respect and humility.*

With quantum mechanics it's very tempting to say, "Because electron A and B became entangled, when we take A and send it away and change its spin, B instantaneously changes its spin." This means there is some kind of information field. That would explain why we could do prayer from a distance or telepathy or precognition. But physicists like you warn that no, you can't say with any validity that non-local entanglement explains psi phenomena. I would say that criticism is valid to a certain extent. We have to be careful when we make scientific statements not to go overboard. In all my writings, I point this is where science goes and this is where it stops. There are meta-laws, universal laws beyond laws. Basically, three is the number associated with the goddesses in the Hindu traditions. The triadic nature of consciousness is in fact how consciousness manifests. But three of those laws are more particularly obvious in quantum mechanics. I call them "quantum-like," meaning they apply at all levels of reality.

What are the three laws? The first one is complementarity and the second one is universality that allows knowledge to arise. If we didn't have universality, we couldn't understand each other. What is behind the universality is that the reality that applies to you also applies to me. Scientists take for granted the second law which is the law of universality. We can find patterns like the spiral pattern that is a dominant pattern in the universe. It exists in spiral galaxies, in dynamical systems and in plants with wonderful spiral patterns. The third law is interactivity in that everything in the universe interacts with something else. Even consciousness interacts with itself. The laws apply not just to quantum physics but also to classical physics, which is particularly obvious in quantum physics. This is where the term "quantum-like" comes from.

Quantum-like encompasses quantum at the micro level, but extends to large dimensions. For example, the sensory systems of the human being are quantum. There are Buddhist monks who can detect single photons. With training, they can see a single particle of light, quantum phenomena. They can also detect a huge number of smells, which can only be explained in terms of quantum phenomena. We have quantum processes in the brain. So, the question is, when we say the particles in two opposite directions are connected, can that explain psi phenomena? The answer is no because we are referring to individual particles. However, in terms of these meta-laws or in terms of the quantum-like the answer is yes because of complementarity.

Complementarity is the foundation principle in the quantum physics of Niels Bohr. He said it applies to all different systems. I'm working with biologists to look at this interaction between biology and physics (*like Nobel Prize-winner Brian Josephson**). You can say photons are non-living, whereas biology is a science of living. But it's not a sharp boundary. The boundary extends through the three laws that are the foundation of the universe. Instead of complementarity, you can use entanglement, which is easier to be understood by a large number of people because after all, science has to reach the people. Entanglement is a little difficult to understand unless you talk about a mother and a child relationship that is always entangled.

In terms of the conscious universe, could you explain what complementarity means in terms of Bell's theorem? Complementarity is the wave-particle aspect. In terms of Bell's theorem the reason that there is entanglement is because of the basic wave nature of quanta. So when you make a measurement, it shows up as a particle but in fact it is the wave aspect that connects them together. If you consider quanta as waves it's easy to understand; otherwise it's an impossible problem. *How do these entangled particles, photons, electrons, communicate?* In some of the recent papers they would say it is the information field, communicating through non-space-time.

Do they communicate with waves or is it something totally different? They communicate instantaneously because information is instantaneous. This is why the universe is mind-like. *But what does information field mean? Can it be measured? Does it have any properties or is it like an invisible vacuum that somehow magically conveys information?* It's not magic because the universe is non-local; this is what Dean Radin has been trying to say. You can call it magic if you follow the Cartesian universe and it makes no sense and is rejected. It is information because information comes from two words, *in formation.* So intelligence or information means that you can make sense of random chaos from orderly information. This is what the human mind does very well and why we have computers that are like us. Thus the universe is informational in nature or conscious.

We can measure our brainwaves--alpha, beta, theta, delta--but is there any kind of measurement of the information field? When particles are communicating it's through the information field. The three-part system of consciousness is being/existence, ability to know, and ability to create an infinity of universes. Whenever we talk about consciousness we get in trouble because we think of human consciousness rather than the more fundamental level. I think the universe is informational in the sense of the cosmic mind, not in the sense of a human mind. *So it can't be measured?* The entire thing does the measuring and the entire universe is the observer. Observing means information. *It's an awareness but it's not tangible?* It's fundamental awareness. You can't prove existence. Existence is existence. Second is the ability to know self, consciousness, and the third is the ability to create everything with the power of the mind.

Our concept of time is linear, which we know isn't accurate because of Dean Radin's experiments and the remote viewers who can view a target before it's selected. Exactly. If anything, time is circular. *We're so used to thinking time is linear.* That's because we have a linear mind. *But then you have to look at it, it's linear only in terms of where it is on the circle.* Actually it's a spiral. *I wrote a book for kids called* Answers to Kids' Deep Questions in Photos. *For the question, "What is God?" I include photographs of spirals in galaxies, a rose, and a nautilus shell.* This is the second law.

What's the theme of You Are the Universe: Discovering Your Cosmic Self and Why It Matters *published in 2017?* Basically Deepak Chopra and I ask fundamental questions about the universe. "You Are the Universe" means you are universal consciousness so that the separation between observer and subject is not real. We are what we observe.

Scientists haven't succeeded in formulating a theory to unify the physics of the big world and subatomic world. Do you think that these three laws in a way unify them? Bernardo Kastrup* is a very good thinker in the Eastern tradition but he is also a Western computer scientist. We wrote a paper where we said basically consciousness is the big white elephant in the room. I think this search will be futile until it's acknowledged. *Why do you think that academics are so afraid of consciousness? Wikipedia makes a point of saying that someone like Radin does pseudoscience. Why are the skeptics so emotional?* It's emotional and people start religious wars over it. In a sense, a lot of the so-called scientists are religious zealots. Their ancestors burned witches because they were afraid of the feminine aspects.

I talk a lot to scientists and find there are a lot more of us out there than meets the eye. Young scientists are concerned because you don't get tenure if you start doing woo-woo sciences. I was actually also careful before I got tenured at George Mason not to cross that line, although I was teaching a course where I used meditation as a first step to study physics. The students loved it. I don't reply to people who are nasty to me. I just ignore it completely because once you get into the game, the third law in terms of interacting on the same level with nastiness, then you have religious wars.

Your book Looking In, Seeing Out: Consciousness and Cosmos *was published in 1991. Was that your first metaphysical book?* Yes, as the book states, science and spirituality don't have to be opposite. (*You like women with Ph.D.s.* It's my type. They get me thinking.) Descartes taught us the mathematical universe is clockwork, but Newton was a great mystic, a natural philosopher. *You and Robert Nadeau wrote* The Non-Local Universe: The New Physics and Matters of the Mind *in 1999 where you talk about experiments in terms of entangled photons.* Nadeau was in the English

Department and got criticized for working with scientists, but he was an expert on Niels Bohr. That's how I got interested in Bohr's Copenhagen interpretation of quantum physics. *The Conscious Universe* that was written with Robert Nadeau and then we wrote *The Non-Local Universe.*

The idea is that entanglement shows us that there is no time, no space? In the book you say that this is the most momentous discovery in the history of science. Yes, I still believe it is. It's too bad that John Bell died early because he certainly deserved a Nobel Prize for the most momentous physics understanding that our universe is non-local. The EPR paradox was rolled out by Einstein-Podolsky-Rosen but actually Bell asserted it earlier. We think of the universe as made up of objects but if it's non-local then objects are an idea in your mind. I believe the wrong understanding can hold us back if we believe we are something less than we really are.

I was taught that an atom is like little balls rotating around a nucleus. But now we know that atoms are potential vibratory vortexes. They don't exactly exist. The staggering fact of quantum physics is that they're waves of probability. In other words, they're in your mind. The particles are in the mind, so they're not hard little spheres that go around like the Earth goes around the sun. They're waves but when you measure them they become immortalized. *Do the waves exist in some material way? What does potentiality mean really? I know intellectually that 99% of you is empty and it's all vibrating potentiality. But why can I touch your shoulder and feel solid?* It's not 99%, it's 99.9999999% so that most of our body is empty space. Because of magnetic interaction, electricity, and atomic forces you can't go through my skin if you poke me because there's resistance from chemical bonds and forces. Some healers can go through your skin without an operation. *Barbara Brennan wrote in* Hands of Light *and* Light Emerging *that her guides would go through her fingers and do psychic surgery on organs.* We have to be careful because we are also scientists but at the end of the day the body is not solid. It's fluid.

Placebo effect has been an irritant to the medical profession and the drug manufacturers but they should say, "Placebo is 40% as effective. Why don't I learn how to harness that?" Then we wouldn't need to use pharmaceuticals with side effects. Sometimes the placebo works better than the pill. On the other hand, we don't want to throw out the great advances of Western medicine because it's doing phenomenal work. But certainly sometimes a placebo works better than the chemical that may poison your body. *Why does my mind believing, "This big red pill is going to help me," change my pain level or my migraine, even if I know it's a sugar pill?* Because the universe is mental and conscious.

Someone may say, "Sometimes believing in getting well doesn't work," and you have to have an operation; I had an operation myself. Sometimes there are deep ingrained conditions that some people say come from karma or past lives. *Ian Stevenson found about one-third of children have birthmarks or phobias that represent the way they died in a previous life. If karma can even imprint the physical body then it certainly implies that we carry imprints in our emotional bodies.* Exactly. The emotional body is much stronger than the physical body. There is a continuum between the physical and emotional and biotic like when we dream our body expands, it is the energy body, beyond the physical body.

What do you think will happen when you die? People always ask that question and the number seven rule is that one day we're going to die. The real question is, "Where do you come from?" If you know where you came from, then you know where you're going. In fact we know from Tibetan Buddhism and some of the Hindu teachings about what happens when one leaves the body behind. *Like the* Tibetan Book of the Dead, *going through the Bardo, the different stages?* Exactly. The Christian tradition also teaches that the soul lives on—it's universal. My question is what brought me here?

Do you remember previous lives? I can't say whether it was previous lives or being in a strong emotional situation like an area where a lot of people were killed. For example, when I visited Waterloo I wanted to get out. A few thousand people got killed there in one day. *With Napoleon?* Yes, so that energy is still there. Now was I there? Maybe I was picking up the negative energy that was still there. As a physicist I can say there's really no way to tell the difference between previous lives and parallel lives. Maybe we are tapping into a parallel universe. If the universe is conscious then the whole thing is virtual. A Korean self-taught Zen master named Daehaeng Sunim was very intuitive. She taught in some of her Dharma talks that when she visited Germany she started feeling all these lost souls. They informed her that they were priests and nuns brutally killed during religious wars between the Catholics and Protestants and they were stuck. They asked for help, so she urged them to go on and they did.

Why do you think that the world is in such a bad state? If we're part of universal consciousness why are we so corrupt? Because of linear thinking, because we want to dominate each other and in dominating each other we probably kill each other. This is the lesson we haven't learned from history so we keep repeating the same things again. Einstein said, "I can tell you World War IV is going to be fought with sticks and knives." My concern now is we are at the stage where we can stop this for ourselves. *Do you have hope about our future survival?* Absolute hope. Yes, because the universe is conscious. Life is everywhere in the universe so if we fail here, we'll pick up somewhere else because there's one followed by 22 zeroes of Earth-like planets in the universe.

What's your next book? What are you researching now? You've been doing a lot of work about climate change. We publish mostly in that and we do some work on natural hazards. My current book is *Science, Reality in Everyday Life*: Science, reality, which is metaphysics and philosophy, and everyday life which is the eighth piece of substance. I apply quantum mechanics to everyday life. The second part has to do with philosophy.

You've written over 330 articles and 20 books; what do you do for amusement? I hang around people and we talk. I go to dinner.

Is there anything else that you would like people to think about? Do hard work. Don't give up and remember you are more than you think you are. Follow your own life. *Follow your bliss.* Yes and don't listen to what other people tell you.

Books and Monographs

Astronomy Laboratory Manual, 1984

Kafatos, M., Harrington, R.S., Maran, S.P. *Astrophysics of Brown Dwarfs*, 1986

(Ed.) *Supermassive Black Holes*, 1988

(Ed.) *Supernova 1987A in the Large Magellanic Cloud*, 1987

(Ed.) *Bell's Theorem, Quantum Theory and Conceptions of the Universe*, 1987

Kafatos, M. and Kafatou, T. *Looking In, Seeing Out: Consciousness and Cosmos*, 1991

(Eds.) Kafatos, M. and Kondo, Y. *Examining the Big Bang and Diffuse Background Radiations*, 1996

Nadeau, R. and Kafatos, M. *The Non-Local Universe: The New Physics and Matters of the Mind*, 1999

Kafatos, M. and Nadeau, R. *The Conscious Universe: Parts and Wholes in Physical Reality*, 2000

Kafatos, M. and Draganescu, M. *Principles of Integrative Science*, 2003

(Eds.) Zhou, G., Baysal, O., Kafatos, M. and Yang, R. *Real-time Information Technology for Future Intelligent Earth Observing Satellites*, 2003

(Eds.) Qu, J.J. et al. *Earth Science Satellite Remote Sensing: Science and Instruments, Vol. I* and *Data, Computational Processing, and Tools, Vol. II*, 2006

Chopra, D. et al. *Cosmology of Consciousness: Quantum Physics &*

Neuroscience of Mind, 2011
Kafatos, M.C., Chopra, D. *The Time Machine of Consciousness: Quantum Physics of Mind*, 2014
Chopra, D., Kafatos, M.C. *You Are the Universe*, 2017
Living the Living Presence, 2017

Ruth Kastner, Ph.D.

The Transactional Interpretation of Quantum Mechanics

Photographer: Anonymous

Questions to Ponder

Dr. Kastner maintains that some theories about quantum mechanics deny free will, since according to those theories, all actions are predetermined. Her preferred Transactional Interpretation allows for free will. Explain.

The Transactional Interpretation suggests there's a dimension beyond space-time, more fundamental than the phenomenal world. Explain.

The concepts of dark energy and dark matter are not descriptive or accurate in her theory, according to which the phenomena attributed to them arise from the structure of space-time. Comment.

When and where were you born, and what are your astrological signs and Myers-Briggs profiles? I was born in 1955 in Schenectady, New York, the second of three children. I'm a Libra and ENTP. My mother majored in physics at McGill University in Montreal, so she was a pioneer in that area. My father also was a physicist, as were several uncles. My mother was told by the director of the lab that women would not be allowed in the physics lab. When lab class began, she went in anyway, basically daring him to throw her out. He didn't say anything, so she continued on. That's the kind of person she is. She's now a retired math educator. I did encounter some issues as a physics graduate student. I hate to play the gender card, but there were some double standards but it's steadily improving. *It has to because women are the majority of undergraduates. There was that uproar over the Barbie doll that was programmed to say, "Math class is tough."* Yes, I wrote an op-ed about that.

What was your major as an undergraduate? I started out majoring in studio art because I was into pottery, then I switched to music education. I majored in strings (violin) but developed hand problems so I had to abandon that for a while. I started getting interested in the sciences and in physics and got another undergraduate degree in physics. *What made you think about it other than your family history? That's a big jump from art to physics.* I always wanted to know what is the nature of reality and first thought I would pursue that through art. Then I got involved in the TM movement, Transcendental Meditation. I saw interesting videos of some professors at Maharishi International University. Larry Domash was a physicist and president of the university who talked about physics and quantum theory and quarks––it just grabbed me, so I got into science through the TM movement. I did the TM Siddhi program, which involved training in the yoga sutras. I'm a bit of a lapsed meditator now, but I do meditate occasionally.

What was your family religious orientation when you were growing up? My heritage is Jewish, but we were very agnostic. My father was kind of a rebel and didn't follow in the orthodoxy of his family upbringing. My parents were scientifically minded and didn't think that spirituality was rational. *What was their reaction when you started doing TM?* They were dubious but probably thought, "She's young so she's going to do crazy things." *Many scientists I've interviewed have a Jewish background. Your thoughts on why Einstein, Freud, Marx—many thought leaders—–come from a Jewish tradition?* There's a strong cultural value of education in Jewish tradition. *The same is true for Asian families but I don't have an Asian visionary scientist so there's something else besides valuing scholarship.* My sense is that in Asian cultures there's a stronger drive towards a collectivism in thought, more obedience to authority, so people may feel, "I don't want to question. I want to work hard and learn and value achievement." As far as going off the beaten path, perhaps there's not as much freedom to do that in that culture. *Right, because in their countries of origin they'd been the dominant culture which isn't true for people who have a Jewish background. They've had to go around the dominant culture.* Right, they're marginalized so, "I have nothing to lose in exploring these obscure ideas."

What had been your most difficult challenges and how did you cope? My challenge is developing the model I've been working with. John Cramer's Transactional Interpretation grabbed me like an epiphany that when I was finishing my Ph.D. in philosophy of physics. It's a way to make sense of quantum theory (QT), but I've had to go it alone because it isn't a popular way of approaching the interpretational issues of QT. I'm swimming upstream. The interpretation has been rejected based mainly on metaphysical prejudices. One of the problems I saw in interpreting QT was that it was limited to the non-relativistic range, which is not the full range of nature. Nature is relativistic such as at higher speeds and energies. Relativistic physics is very different from non-relativistic physics. It took lot of self-teaching even though I have a master's in physics and did some graduate work beyond.

Being a faculty member, author, wife and mother of two daughters is a lot to juggle. I wasn't employed full-time while my daughters were young, so I was at home. Fortunately my husband was an electronics engineer, able to carry the financial burden. Also, because I've been working on an area that's viewed with dubiousness by the academic community, I've never had a steady full-time academic position. I'm an independent scholar in many ways.

Let's talk about this transactional interpretation of QT. An atom produces an offer wave to indicate, "I have a particle I don't need anymore. Does anybody want it?" When you have an atom in an excited state where one of its electrons has an extra quantity of energy, it will take the form of a photon, a quantum of light. An atom in this excited state is a bit unstable and wants to relax down to a lower state. When we talk about offers, at the relativistic level it's not really a unilateral thing where this one atom is saying, "I've got an offer, how much am I bid for this photon?" There's a mutuality to it so that an excited atom will not be able to generate this offer unless there are potential absorbing atoms around. These are atoms in a lower state, like a ground state in which they would be capable of assimilating the photon and popping up to a higher state.

You have to satisfy energy and momentum conservation in this process, in a kind of non-locality that people find challenging, in what's called Direct Action Theory. It's a specific theory of fields that the transactional picture is based on, in direct contact through virtual connections and virtual photons. They're not offers or confirmations; they're just a non-local awareness that these atoms have, so they kind of know about each other. When you have an excited atom and some other atoms in lower levels, the atoms communicate mutuality. At some point, and this is indeterministic, they can generate an offer and usually many different confirmations from different possible absorbing atoms in a handshake process.

Do you know that by observation or experiments? Or is it a mathematical model? I'm thinking of it like a dog in heat when the other dogs sense she's in heat and try to mate with her. That's a good analogy for the basic idea here. It is a mathematical model that is not something that anyone can detect experimentally. It's about non-locality and sensing this is a property of the Direct Action Theory of fields invented back in the 1940s by John Wheeler and Richard Feynman. They tried to deal with problems in physics having to do with infinite self-energies where a charge seemingly interacts with its own field. But the usual theory predicts enormous energies that we in fact don't see. They developed the Direct Action Theory as a way of getting around that problem. It accounted for the fact that when you have an accelerating charge it radiates and loses energy. In the usual approach to electromagnetic theory it's hard to explain that. Later they abandoned the Direct Action Theory because they felt that they didn't need it and it fell into disfavor. But it's a perfectly good model and it does have this advantage of explaining certain things that the standard approach has trouble with. It's a mathematical model that explains certain theoretical problems better than the standard approach.

So the atom sends out an offer, "I've got an extra photon," and then an absorber says, "Okay, I need it." The particle is transferred but it happens retroactively? Please explain the time lag. There's a lot of confusion about this, possibly because there are different ways to understand what's going on with this so-called retroaction or retrocausality. People usually think that these confirmations are advanced waves that go back in time. My proposal deviates from that picture in that these offers and confirmations are taken as possibilities not in space-time. Even though nominally these waves have a forward-directed or backward-directed temporal character, I view them as possibilities so that they are not actual space-time processes. They are possibilities that can give rise to different space-time events.

We have competing absorbers, sending advanced confirmations but they can't all receive one photon. So they're all in play as possible recipients but only one of them will win the photon. At that point, you're generating a new space-time interval, creating a set of a pair of space-time events that literally did not exist before. It's not that you have a signal going back in time but you established a new space-time relationship where, from the standpoint of that absorbing atom, an absorption event is like the "now." If that atom were conscious it could say, "I have now received this photon in a space-time event in my now but its emission is in my past." The emission event will be established toward the past of that absorption because something has to be emitted before it can be absorbed. If you think of it as a knitting process, you have a stitch on the needle––the absorption––and the previous row is where it was emitted from. That is a new space-time event, so we're not sending influences or signals back in space-time, we're creating new elements of space and time.

You talk about different levels of reality like the middle ground that's not as real as an actual event. How does that work? This gets back to your interest since you were a girl in what is reality. QT points us to aspects of reality that are veiled or more subtle, and these are the possibilities. It's instructing us that reality has greater content than we thought, in that there's more going on than just space-time. In my books I use the metaphor of the tip of the iceberg. We usually think, "Everything that's real exists in space and time," but what I'm saying is no, space-time is only the tip of the iceberg. There's a vast submerged aspect to reality that exists, even though we can't detect it with our five senses. Quantum theory suggests that possibilities are more fluid. You can't have one photon going in different momentum directions. There are far more possibilities than the actuality we experience.

Do you think the substrata existed before the Big Bang? It could have been the first thing to come into being; you have to have possibilities before you can have actualities. *Does this relate to the fact that we only understand 5% of matter and all the rest is dark energy and dark matter that we don't really know what it is?* In the Transactional approach there's a way to understand what are called "dark energy" and "dark matter" as properties of the emergence of space-time. I wrote a paper with Stu Kauffman explaining that if you look at the way space-time is generated, it has properties that will mathematically give you the kinds of effects that we attribute to this missing matter. It's not really that there's invisible matter out there, it's that space-time has a kind of structure that leads to the phenomena that we're seeing. It's a hard sell because most people think reality is in space-time. When you say it's a structure emerging from something more fundamental, there's a lot of dubiousness about it.

If we go back to our dog metaphor, we have a dog who is not in heat but she could be and somehow the male dogs start being aware of this before she goes into heat, what makes her go into heat (be willing to release a photon)? At the level of the math, all you're going to get are probabilities. In my model it's intrinsically indeterministic, which means that there's nothing in our current physical theory that you could point to that says, "Here's why this happens at this time." If things are not completely deterministic and there is this fundamental indeterminacy, that's an opening for volition to come into play. I've tried to debunk the usual claims that science tells us we don't have free will, which are false claims.[1]

What would Einstein say about this theory? How does it tie in with General and Specific Relativity? He famously said that God doesn't place dice with the universe, so he didn't like this randomness. But if you think of the world as being completely deterministic, then it's very hard to say why anyone would have free will. As far as harmony with relativity, you could get a form of quantum gravity out of this approach. If you think of space-time as emergent from the quantum level, you find that the way that the events emerge are relativistically covariant and uphold the strictures of relativity theory. In his time, people thought of space and time as the whole of reality, so I can't say Einstein would've signed on to a challenge of that metaphysical view.

Could you call the substrate below space-time consciousness? The simplest metaphysics is that mind or mental substance is fundamental because out of that you can get everything. It's a consistent metaphysical view to have a monist metaphysics and say that things are fundamentally Mind. Although that's not something I would go around on a soapbox campaigning for, to me it makes sense.

It's really tempting but I know it's fraught to say, "Entangled particles can influence each other from a distance non-locally, indicating an information field." It could explain why we can do telepathy, clairvoyance, healing from a distance, precognition and retrocausality. Most physicists tend to be empiricists and are loath to make those kinds of inferences. At the quantum level--again these things are not in space-time but rather beneath the surface of the manifest--there is no distance between possibilities. There literally is no distance because there's no space-time. These possibilities can exchange information because they're all at the same place, unseparated until a transaction happens.

In so-called "non-locality" there is no real action at a distance because there is no distance. To me it's a consistent metaphysical view to propose that if consciousness is fundamental then seemingly non-local exchanges of information of the kind that go on between entangled quanta could go on between living beings as well. When people are dogmatically opposed to exploring those possibilities, they're being dogmatic and empiricist. They're making a metaphysical methodological choice to be empiricist, believing they can have no knowledge of something unless it affects their five senses.

Is the world composed of some mental substance or is it made out of physical substance or is it somehow dual? I personally don't think that dualism makes any sense because if you've got two different substances, there's no way for them to interact. That was a famous problem for Descartes. Then if you want to be physicalist and say the basic fundamental substance is dumb matter, then you really have trouble explaining why anything should be conscious, that's the hard problem of consciousness.

Quantum physicists talk about an information field. What are the implications of that and does everyone agree that there's an information field? A lot of physicists think of quantum states as an information field. A lot of people equate QT with information, but they don't literally mean there's a real field out there. A chapter in my *Understanding Our Unseen Reality* book criticizes some of the interpretations that rely on the term "information" because they're very equivocal about what they mean by it. *But information is something. It's not nothing so it has some kind of form.* Sometimes when people use the word "information" they mean gaining knowledge rather than something that exists somewhere in the world. Say there was some accident in New York City but it's only information to me if I change my state of knowledge. They don't mean that information is a field that exists somewhere in the world.

But it's still something since thought transfer and information transfer doesn't happen in a void. Yes, you could have a signal beam that contains that information. But some people think of the quantum state as not referring to something in the world but simply being an indication of what I know and what I don't know. That's the anti-realist use of quantum state as information. I criticize thinking of all space-time as just all the results of measurements, out there in a block world from the beginning of time to the end of time, so it's already decided in the universe where each photon is going to go and what the measurement results are going to be. This is what a lot of people think right now.

In that picture, when I assign a quantum state to something, the quantum state doesn't tell me which outcome will be realized. It's like an incomplete descriptor, just a measure of what I happen to not know at this particular time. That's a very impoverished use of the term "information." It's a measure of partial information or a measure of ignorance. It's very observer-dependent for a particular observer because some other observer in the future will have more information and will assign a different quantum state that's more focused to a particular property. In that view people are just using quantum states as labels for what they happen to know at a particular time.

In the double-slit experiment, if it's not observed, the pattern will be different than if someone is observing or measuring the photons going through the slits. Is that because it's determined that it'll go through the single slit or the double slit? In the Transactional picture you explain it in terms of absorbers and confirmations, which is different from the so-called "epistemic" view of the quantum state as just a measure of one's ignorance at the time. According to the epistemic view, there are specific events already in the future where a particular kind of measurement or absorption or detection will be performed in that experiment and there will be some outcome although that person doesn't know it yet. That's their model. *So it is deterministic?* It's not deterministic in the sense that the quantum law doesn't specify which will happen, so they think there are discontinuities because it doesn't tell you where the photon is going to end up. There's this brute force discontinuity but something happens, one of the outcomes just happens according to the probability that quantum theory lays out and that's in the world already, but we just don't know it. They think of observers as moving through this block world and assume that all measurements take place to the end of time even though the law does not determine which result occurs.

How would you explain that observation changes the way photons go through those slits? In the Transactional picture, observation is not something that we have to invoke, since it's about the responses of absorbers, which may or may not be made out of something that's conscious. When you have these absorber responses you have a collapse situation analogous to spontaneous symmetry-breaking. Measurement is the interplay between emitters and absorbers. *How does observing it or measuring it change the outcome?* Without absorber response you won't get any outcome. So if we think of our excited atom, there's no measurement, there's nothing happening, then at some point they can mutually decide, "We're going to transact." That means our measurement is going to happen. That's the first stage of measurement and at the second stage one absorber has got to win and the rest will lose in the symmetry-breaking stage. We can think of that as an opening for volition. It's not that observing changes the outcome, it's that it's a process of measurement on this level of mutuality where once a measurement situation is set up, only one thing can happen.

Is it possible that the human observer or measurer could influence what atom accepts the excited proton? It's possible. There's nothing in the theory that provides any mechanism for that, but it's possible. This is where you get to the principle of sufficient reason, you get into these questions about whether anything can happen if there's no specific reason for it. The answer "no" is known as Curie's Principle because Pierre was of the view that absent a particular precipitating cause, nothing would happen. We really don't know why one outcome happens as opposed to the others.

There are many studies of precognition and retrocausality and other psi phenomena, so if you were pressed, what would you say explains them? I think the jury's still out on whether the results that people have reported from these kinds of experiments are genuine positives or false positives. I have to caution about experiments that have been presented as having conclusively shown these things because there are some troubling findings, such as Daryl Bem's article on precognition. Psychology and sociology researchers found that the field had not been aware of how easy it is to generate false positives when people think that they found an effect, but that because of other variables they didn't control for, it may not be a real positive. But there's nothing in the science that can say, "These paranormal experiences aren't true."

The psi findings are instead of 25% by chance, a hit rate of 27% or 33%. This gets back to the way these experiments are often done where it turns out you can get that apparent effect but it's a false positive so that the level of significance may not be a real positive effect. I was at a conference in 2011 where Daryl Bem presented his results but later it turned out that the entire field was following procedures that turned out to lead to a lot of false positives.[2] A paper came out showing that you can show that people get younger by listening to the Beatles; by having certain variables that they didn't control for carefully, they got an absurd result.[3]

Let's talk about your books. The Transactional Interpretation of Quantum Mechanics: The Reality of Possibility *was published in 2015 and 2017. What was your main point in that book?* To try to get people to consider the possibility that there's more to reality than space-time and QT refers to this field of possibilities. You need to think outside the space-time box. *How would you define consciousness? Is an amoeba conscious?* You could say that the basic feature of consciousness is the ability to be aware. So the fact that we each have our own subjective sense of being aware of something is a basic experience that in some sense is self-empirical. You have a field of awareness which is like the intransitive aspect of it but then there's the transitive form of it which is being aware of something.

What's your personal feeling about what happens after you die? Do you think your consciousness goes on? I sure don't know but I've read some channeled literature by Paul Selig and a book ostensibly written by Lord Maitreya channeled through Kim Michaels. It would be interesting to have a real paranormal experience someday.

Understanding Our Unseen Reality: Solving Quantum Riddles *was published in 2015.* This book is more for the layperson and I used illustrations to convey the ideas. *Your newest one is* Adventures in Quantumland: Exploring Our Unseen Reality. In that book, I was answering the question, "What do you mean we create space-time out of these transactions?" The ancient Greeks were brilliant, but they also had some ideas that have constrained us, like the static eternalist idea that the whole world is out there already or that the whole space-time construct is out there. There was a physicalist idea, the atoms in the void. What I wanted to do was counter that and say, "Not all the Greeks thought that way. There was more of a more dynamical aspect to their thought too." I talked about free will because of some of the negative statements coming out of a lot of the physics community and the neuroscience community, very dogmatic categorical denials of free will. There are real consequences that come from convincing people they have no free will. *Hindus and Buddhists might say that our karma set up by past lives predisposes us to have certain patterns in this lifetime but we have choices to work through the karma or not.* Saying that you have free will doesn't mean you're not influenced and that we don't have enormous constraints on us.

My other book, *Quantum Structural Studies,* is an edited collection about the classical or space-time emergence from the quantum level. My hobby is Tudor music in Renaissance England, so I also wrote a screenplay and book titled *The Harmony of That Heavenly City* with my sister Judith Skillman, about William Byrd who was the court composer for Queen Elizabeth, because I love his music. We wrote a screenplay about his amazing life that was reflective of the plight of the Catholics during the Reformation. Byrd was a Catholic working for a Protestant Queen. He did very shady deeds, hanging out with Jesuits and traitors while the Queen looked the other way.

It's fascinating cloak and dagger stuff.

In the face of climate change, growing inequality, and the rise of autocrats, are you optimistic or pessimistic about our future? I try to remain optimistic. It is scary and challenging. I think we're at a pivotal moment and we've got to push through and make a lot of changes.

Anything else that we should include? The science versus religion and spirituality issue needs to be corrected. Many people think of them as at each other's throats although really they're very complementary. I write about that in the final chapter of my latest book *Adventures* where I say, "We can reconcile the scientific and the spiritual traditions. They're different ways of knowing. Each has its limitations but each has its power and they need to work together more closely." We have inner ways of knowing that are legitimate, but scientists would scoff, because they're not empirical. If we take QT seriously, it's pointing us to an aspect of reality that is sub-empirical. Human beings are way beyond computers, logical systems, the rational. The rational is always limited in that Kurt Gödel's Incompleteness Theorem points out that every logical system has its limitations. So science has to accept its limitations and not try to judge other ways of knowing based on standards that don't apply.

Books

The Transactional Interpretation of Quantum Mechanics: The Reality of Possibility, 2012

Understanding Our Unseen Reality: Solving Quantum Riddles, 2015

Quantum Structural Studies: Classical Emergence from the Quantum Level, 2016

The Harmony of That Heavenly City, 2017

Adventures in Quantumland: Exploring Our Unseen Reality, 2019

Endnotes

1 Kimberly Anne Clinch. "The Physics of Free Will." Helix:

Connecting Science to You, January 14, 2018. https://helix.northwestern.edu/article/physics-free-will

2 https://www.ncbi.nlm.nih.gov/pmc/articles/PMC4706048/

3 https://www.theguardian.com/science/2012/mar/15/precognition-studies-curse-failed-replications
https://journals.sagepub.com/doi/full/10.1177/0956797611417632

Fred Alan Wolf, Ph.D.

Is There an "Out There" Out There?

Photo used by permission

Questions to Ponder

Dr. Wolf concludes that mind is primary. What is his reasoning?

How does he explain quantum non-local aspects of entangled particles?

Dr. Wolf warns against thinking in boxes. Explain.

How does the author view time?

I was born in Chicago, Illinois, December 3, 1934. *Do you identify with being a Sagittarian?* What does it mean to identify? I recognize that under the terminology associated with astrology, I was born under the sign of Sagittarius but the word identify is too much to describe. To identify with being a white American, that's too much. The word "identify" has a kind of a closure to it. It puts things into too tight of a perspective. I recognize and I understand the principles of astrology, so yes, I was born under the sign of Sagittarius.

You've taught and traveled all around the world. That spirit of adventure and travel is a Sag trait, right? That's what I am told. But I would imagine there are other people who are not Sagittarius and are just as adventurous as I am. *Like me. During your travels, do you find that you teach differently when you are teaching in London or San Diego or…?* Absolutely. One universal principle I have when working with an audience is to test the audience to see how they are responding with certain phrases or terminology and to encourage questions. Through that interaction, there is a tailoring that's automatically going on. It's just like when you are speaking to kids in kindergarten, you wouldn't give the same talk that you would give to graduate students in physics. *My experience is that there are differences. I've taught a lot of workshops in Japan and I had to work hard to get them to interact and ask questions but they would open up.*

What kind of family did you grow up in Chicago that made you curious and want to be a scientist? Nothing made me be what I am. You might say what was allowed is a type of pruning of what I am not. Family is not there to make you be something but to prune away the things that get in the way of your achieving. I was pretty much a bright kid. I had emotional problems. I stammered as a child and my mother worked very hard to get me through that period. My father was more or less the disciplinarian, but he didn't really spend that much time with family and raising me. I would say family let loose or allowed me to be nurtured, to become what I still am.

What's an example of something that your mother pruned that allowed you to excel? Because I was stammering and had a problem, my mother looked to see what it was she could do to get me to speak better, helping me move into a direction which should be better for me. I think my parents, mostly my mother, recognized that I was a fairly intelligent and curious child because I was always taking things apart and putting things back together.

Do you think that the focus on learning to speak played into you becoming a public speaker? Oh, absolutely that was a major influence. If I hadn't been a stammerer, I never would ever have gotten into public speaking or any of the things I do now. A couple of things that I learned as a child which I think helped: my mother had an interest in music and I developed that interest too. My mother and uncle brought me records of classical music and I became informed. In my stammering period, I wanted to play harmonica. At that time the greatest harmonica player in the world was Larry Adler. He had a book on how to play the chromatic harmonica and I learned how to play it, which improved my breathing technique, so that helped with my stammering.

I also was interested in magic, which might have been a good influence on my writing because I am still interested in the magical experiential quality. Even though I am a quantum physicist, I still have an interest in what makes human thought. I liked close-up magic rather than stage magic, mainly because I can work with a deck of cards. When you are working with close-up magic, you must talk. In order to learn how to patter, I had to learn how to speak. Magic is very much a part of the way I do my speaking engagements, using PowerPoint presentations with animation. I don't just get up there and speak, I show things. So, this all came about from my early childhood. *It's interesting that Russell Targ was also a magician as a teen, both of you thinking outside the box.*

What about your family's religious or spiritual teachings when you were growing up? I am a Jew, but we never actually practiced Judaism in the house. There wasn't any emphasis on that as my parents were first generation Americans because my grandparents came over from the Russian ghettos. I did go to Hebrew school and I learned some Hebrew and I picked up on some of the ideas in Judaism, but it was really not an important part of my life. The temple that we went to was Reformed, which is about as gentile a Jew as you can become.

How did you get from Chicago to UCLA grad school? When I went to high school, I had two major interests--girls and football. Males between the ages of 13 and 18 are very much penis-minded. They can't help it. I was a quarterback in high school football. I took a physics course and it was very dull and boring. I had my pick of universities, but due to financial requirements, I went to the University of Illinois and got my degree in engineering physics. Engineering because I like to take things apart and put them back together again, and physics because I wanted to know how this universe really works.

By the time I graduated from college, I was an ROTC cadet, so I became commissioned as a Second Lieutenant in the Air Force because we were in the Korean War and the draft was still a part of life. Then I got married and my wife and I moved to Los Angeles because I received a fellowship from Howard Hughes. I was allowed to pick a school--either Caltech or UCLA. I was very uncertain of my abilities, very shy, almost timid. I wasn't really confident that I was smart enough, so I didn't choose Caltech, the major institute on the West Coast; I chose UCLA. I am very glad I did because I felt comfortable there. I also worked part-time for Howard Hughes Aircraft Company in Culver City and had two children by the time I got my Ph.D.

Did your wife help you become more confident? No, during that time, that wasn't part of the marriage contract, so to speak. It was more a question of agreements and we agreed that we were going to get married and my wife wanted to have four children, so we started raising kids. There wasn't time for her to build me up. She had to take care of the four kids. I've got five but that's another story.

Where did the fifth child come from? We've been talking about the years between 1957 and 1962 when I did my Ph.D. Now to go to the fifth kid, we have to go to 1974 in London where I had a research fellowship at the University of London and visiting professorship at the University of Paris. I had been divorced from my first wife and with this wonderful woman I met in London we produced a child. And that child is grown up and has a family of his own and we recently visited.

Who's we? My current wife, my third wife. You've got to understand I've been through many generations. I am 85. I experienced the free love movement, the hippy generation, the beat generation--I've been through a lot of stuff. When I was growing up TV hadn't been invented yet. The first television set, the RCA Victor set, we first had in our home when I was nine or ten years old.

For people who would like to be as active and bright as you are at 85, what advice would you give them based on what's worked for you? I think the best thing is to find joy and happiness in what you do. I like to take it easy, I like to watch dramas on television. I like to read technical scientific papers and work out equations because I want to find out how certain ideas work. I enjoy being with my wife and having conversations about political situations. She is a very intelligent gal and she loves to talk about politics and we get into these really good discussions. I also get down: If something is sad, I'll feel sad.

What else do you do for fun? I breathe. I enjoy life. I do some mild exercising. I used to be a runner and an athlete, so for me exercising has been a part of my life ever since I was a kid. I take long walks almost every day. I don't know what you mean by fun. For me, fun is life and I don't see any difference between the two.

What do you think happens after death? One thinks about it more as we get older. I've had some very deep experiences with death, and because of them I have more of a spiritual understanding of what happens after you die. These experiences started in the 1970s. I used to have lucid dream states, and in one of my states I was with people who had died, who hadn't moved on yet. I could converse with them, as if it was real while they were in a waiting period. It's like these people had committed suicide, and as a result, they were waiting to see how they could best be used in the next cycle of incarnation. From everything I've studied, including my interest in quantum physics, the world is mostly not made of matter; it's made of mind. And the mind is universal. It's one great mind or mind is God, if you want to call it God, but there are different names for it.

Consciousness is an aspect of it. The mind may be primary, consciousness is the action of the mind; mind is noun, consciousness is verb, in my way of thinking. Individuals form limited connections with the universal mind. Mostly their connections have to do with relationships, like bubbles or blobs that form in the ocean. Most of the interactions that go on within these blobs are pockets of consciousness that are capable of reaching out to other pockets of consciousness. As a result, they develop identities. Each blob says, "I," but each one doesn't recognize that they are both bubbles in the same ocean. When you pass on, the bubbles kind of burst and the interconnectedness of the bubble which gave you your likes, your dislikes, your hates and your loves, begins to dissolve. Then you begin to connect with a bigger ocean and then new bubbles form when you come back in again. The process is continually bubbling; it's as if the mind itself has a desire to bubble, to produce these bubbles constantly arising.

It's more of a surrender and then a regrouping. The intelligence and certain basic attributes important for survival, stay intact. I think people become more intelligent and loving as they "evolve." They become more mindful as they evolve because that is what the mind at large tends to put into motion. I would say love is an attribute of mind just as, believe it or not, hate is an attribute of mind. Love is a larger one, maybe, but there are defensive mechanisms that are part of the great mind. There are boundaries between bubbles.

What are the times that have been really difficult for you? My biggest challenge came when I decided to leave my professorship at San Diego State University after I was tenured. I had everything that anybody in their "right" mind would want but I was not in my right mind. I didn't want to be stuck in the position of being a physics professor with blinders that kept you going one direction. I wanted to explore. I resigned although I didn't have another job to go to and I went into the void. I went through a period of you might say depression and finally I turned to God, the great mind. I said, "Look, here I am. I have these talents. I am here to use for the benefit of all. So use me, I surrender, I give myself up to you." As a result, things started to pop into existence and it completely surprised me.

I read that you were the first physicist to develop a cult following in the popular literature. Probably the film What the Bleep Do We Know!? *made you famous?* It certainly launched me into the world's spotlight, but I had already written books. I won the National Book Award for *Taking the Quantum Leap,* so they had a known thing with me. The two people who produced the movie were students of Ramtha. A book I coauthored in the 70s is called *Space-Time and Beyond,* a cartoon book that Ramtha liked and invited me to come to her ranch in Washington. Since then I've made about a dozen different movies.

Ramtha is an entity channeled through J.Z. Knight. Does it seem to you like Ramtha is a figment of her imagination or does it seem like it really is an extraterrestrial intelligence? Both, because what you call your mind is not something that's contained within the material. So therefore, if it is extraterrestrial, meaning outside of the material realm, then Ramtha is outside of that. This entity had some wise things to say.

Did you have other experiences of synchronicity besides lucid dreaming? The worst period of time I had was the death of my son Michael. He was struck by a drunk driver in 1987 and it was a real shock to my system. He came to me in lucid dreams a couple of times and explained to me about parallel realities and said that he was in one and he hadn't really died. When my mother died, she came to me and said that she is teaching, which was to me a funny thing to be doing. She said she was teaching mainly men who had lower position jobs in life to be executives when they are reborn. *Was she an executive?* No, but she was obviously an intelligent woman. These experiences, coupled with my basic quantum physics theorizing, put together a picture that death is not the end, so don't hold on to your personality because that ain't going to last. Whatever you think you are, you aren't. There is something bigger than you are and you will find that they are not going to last, they are just passages.

It's so interesting to me that we understand 5% of the universe, the rest we don't know except to call it dark energy and dark matter. They used to think that most of DNA was junk, but they only know a little bit about our DNA structures. There's so much of the universe that humans don't have a clue about how it works. Yes, we haven't been able to apply a new way of thinking. For example, when we first studied physics, we applied an objective way of thinking. Objects were self-sufficiently "out there," irrespective of mind. With quantum physics, we found that application failed. If we kept using the same approach, we would come up with paradox after paradox. Quantum physics opened up a new door in that somehow the mind is essential to what is going on. You would think that physicists would be jumping on the bandwagon to figure that out, but just the opposite seemed to occur.

One of the things that may explain this predicament is called the "survival mechanism." We hold onto an idea even though it's wrong because we believe it'll help us survive. There are scientists that hold to certain ways of thinking because they believe that's the only correct way to think, but those old ways of thinking will guarantee annihilation.

Like denial of human involvement in climate change. Denial of climate change, racial superiority; a lot of them that have cropped up over the last 100 years that have led to their own downfall because they simply are not within the mindset of what I would call the One Mind. Even my definition of the One Mind is my limited way of dealing with what is experiential and that cannot be really intellectualized; it has to be experienced. Fortunately, I've had some of those experiences in lucid dream states.

A photon doesn't change its potential state until it's observed and then it becomes a wave or a particle? No, it doesn't exist in different states of potentiality; it is a form of mind that manifests through ways that we use to bring manifestation as one form or another; in your terminology, wave or particle. So, it's neither wave nor particle. It doesn't have different states of existence until they are observed and then those states could be counted.

What is it before it's observed? It's part of mind. Physicists use the term "quantum wave function," but it's just a terminology, it doesn't explain it. We can put a mathematical form for it and its structure is very wavy, but it doesn't wave in ordinary space and time; it waves in an imagined infinite dimensional space and time. That is very hard to grasp but that's really the quantum wave function--it's part of the great mind. It doesn't really exist as we commonly think of as things that exist "out there." We deal with it in a limited way in order to explain certain kinds of experiments. There is a principle which allows us to do that. We can narrow things down, we can explore within a certain confine. I put a box around things and only deal with what's taking place within the box, ignoring any influences from outside the box. Physics allows us to do that kind of experimentation.

Let's say there are two photons that become entangled. We send one off to the other side of the world. It changes and the other one changes and responds. What allows for them to communicate? From their point of view, there's no separation. From our point of view, there is a separation in space but from their point of view there is no separation. That's what entanglement is. Imagine you are looking at a very elongated fish tank and there is a fish you can watch from the elongated side or you can look from the end, the smaller side. However, you can only view the tank from the long side, or the end side and you really can only see one window view, end or long, without seeing the other viewpoint. Now if you look through the window of the smaller side (and there is a long fish in there, however, you don't know that), you can see what looks like a fish head growing larger (as it moves towards you) and then you could see it suddenly change to a large tail that grows smaller as you watch, as it grows tinier and tinier until it becomes a dot. Then you see the dot grow as a head growing larger and larger and suddenly change to a tail—a shrinking tail and the scene repeats. First you see a growing head and suddenly it becomes a shrinking tail that changes into a dot that grows and shows a head and so on.

Now suppose your friend is watching the same aquarium from the side view. He sees the fish move towards one end, then swish its tail and return towards the other end and then repeat. Later you and your friend compare notes. You might think maybe there're two different fish in there. There's a fish that changes periodically from head to tail and back again every few seconds just like some fundamental particle changes. Furthermore, your friend tells you he sees a fish swim from one end to the other and turn around, going back and forth periodically. When you compare times, the head-tail change occurs at the same time the fish reaches one end.

You might say the two fish are entangled. There must be two of them in there somehow. But if you look at it from the side and the end at the same time, you are going to see the fish going this way and then turn and go the other way. You see that what looks like separate things is really just one thing. Entanglement is that kind of picture. When there are two or more of them, they can be grouped together so that the observation of one of them also influences the observation of the other. They are separated from a certain point of view, namely the point of view we bring to bear. They are one from another point of view, namely the point of view of the quantum wave itself or the mind of God, however you want to name it. They are not separated in that sense, but they are separated in what appears to be a physical sense.

How does this lead to time? You talk about getting information from the future and parallel universes and time travel. What do we know that puts new perspective on our view of time? I want you to get used to thinking about the language that you use. When you put concepts into certain kinds of boxes, you are asking for answers that fit in those boxes. So linear time, nonlinear time, those are two boxes to you. They seem very different but there's no such thing as linear or nonlinear time because time is not aligned. Take it out of that box and realize that time is part of experience. Now the question is, what kind of experience? There are different ways of experiencing and we can, based upon those experiences, put them in boxes which are linear or nonlinear, but they are neither linear nor nonlinear. Those are the labels we put on the experiences in order to divide them and some way confuse ourselves, fool ourselves into thinking we now understand what time is. We don't. Time cannot be explained linearly, nonlinearly, upside down, right side up, it's neither any of those things.

What do you mean by time travel? If we take a box of time and make a line of time, then time travel refers to moving backwards along the line that's going back in time. Or jumping forward along the line from where you are that's going forward in time. Time travel is simply a way of speaking about moving backwards or forwards from the now present moment along the linear time. Time travel is a way of thinking when we put time in a box. That's it.

Would it be more accurate to think that time is a circle rather than a line? You can make time whatever box you wish to put it in. And in General Relativity, the idea of time being a circle is part of some of the relativistic models, that space-time itself is spherical. There are many different forms, different models of what to make. And once you make time a paraboloid or a circle or a straight line or a wavy line, there are all kinds of models that one can make. Each model is usually used for a mathematical reason in order to explain something about the physics of the experience upon doing certain kinds of experiments. It's not a circle, it's not a line, it's neither of those things, it's really an undefinable quantity that is part of what we call experience. Time and mind are in a certain sense equal.

Dean Radin did experiments where he hooked people up to measurements and they reacted to the arousing slides before the computer even generated them; what allows for that? What do you mean by allows? Is there some forbidding principle that you have in mind that would forbid it? It has experimentally been demonstrated that people can sense things before they happen. *That* allows for that. If I give you a model and I say it's because a=b=c=f and it all goes like that. And you say, "Oh well, that explains it," it explains nothing. It's just words and mathematics and equations; it doesn't explain anything. The fact is that people have precognitive experiences. They have lucid dreams. They experience things before and after time. They have past-life recall. You are looking for a linear model which will convince you that physics will tell you that that is the right answer. No, it doesn't work that way. All the models we use put experiences in boxes and the minute we put something in a box, we say, "Now we understand what that is." Another experiment comes along and blows the box apart.

What about string theory that says that there are 10 dimensions and multi-universes. I've read that the math works but experientially it doesn't. Does that help us understand anything? It helps us put knotty string balls into place, which helps to tie together the gravitational experience with the electromagnetic, the nuclear, the strong, and the weak forces. It brings attempts to bring together all of these different kinds of experiences that could be measured. It has some limited success, but the problem is it can't be measured. We don't have refined measurements that will prove it one way or another. It's a theory that's beyond provability, experimentally. Theoretically it can be done consistently but we can't prove it physically.

You have a concept that that line only goes in one direction, a line with an arrow on it, right? What if there's an arrow on both ends of the line? What do you call that? Nonlinear? You draw a line, but you put arrows on both ends of the line. But it points to the other direction. What do you call that? You could move this way or move that way? It's still linear.

Every human being constantly has information from the future coming into the present. You are born with it; it's called intuition. It's when you sense when something is going to happen before it happens. It's so common and so ignored. What's really important is that time itself cannot be represented. There's no such thing as now; now is a composite of many instances of time from the past, the present and future integrated into your present experience. When you begin to look at things that way, you don't look for the "Oh wow" kind of stuff because you accept that your life is, "Oh wow, gee whiz."

There's a popular belief in the law of attraction, but actually in nature there is a repulsion in that like doesn't attract like. Talking about electrical charges, like charges repel each other. The idea of like attracting like is an idea that's more of a business-type of conceptualization put together by the people that made the film *Law of Attraction*. It never made much sense to me, but it's a teaching tool that basically says that if you really want something, you will attract it to you. My way of thinking is if you really want something, you will move in the direction in which that can be satisfied.

What I think it ignores is that sometimes our best growth is from adversity or not getting what we want or being challenged in some way. It doesn't work that way. It never has and never will.

It's easy to be discouraged and hard to feel optimistic about the world situation today. What do you feel about it? I am optimistic because I know that kind of thinking is wrong. I believe that there is a better way of life and that's not the way. There's always going to be reaction to change, even violent reaction to change. We don't have Germans killing Jews anymore. We are moving in a direction I believe which is actually the true basis on which this country was formed.

What's your next book? What are you studying now? I am looking back at some things I wrote 20 or 30 years ago that have to do with the nature of consciousness, how water molecules work inside the brain and whether water is necessary for consciousness. I am working on an article about the nature of the soul, Buddhist theology and quantum physics. I will put them out free on the Web and see what the response is. If there is a big enough response, I may offer them as chapters of a new book.

Scientists like Dean Radin have shown that psi research has results way beyond probability, that are statistically valid, but we see tremendous resistance. Is that just the same as people insisting that the earth is the center of the solar system. There is always going to be reaction to change. That's called inertia and that's the nature of science.

You've probably gotten criticism from some academic physicists for your books? Yes, it's the nature of the beast. Many people don't like that I popularize or mix in spirituality and physics, and people get really upset with that because they want to stay within their own boxes. Dare I go outside of that physics box and speak about consciousness! Some people are doing that now; the very conservative *Scientific American* is publishing work by Bernardo Kastrup* who goes outside the box.

Quantum physics is definitely changing the size of the box, even opening up holes in the box with more room to breathe than there has been in the past. When I first started writing about time travel, etc. in the 70s, physicists would label us as complete nuts. Now, there are standard articles on time travel. It's a major interest, particularly with quantum computing and the possible use of circuits which go backwards and forwards in time to influence quantum computing. Whatever was forbidden in the past is usually allowed in the present, but new resistance will arise as always. *It's a dialectical process.* Absolutely, it's dialectical.

Books

Taking the Quantum Leap, 1981, 1989

Parallel Universes, 1990

The Eagle's Quest, 1991

The Dreaming Universe, 1994

The Spiritual Universe, 1996, 1999

Mind into Matter, 2001

Matter into Feeling, 2002

The Yoga of Time Travel: How the Mind Can Defeat Time, 2005

Dr. Quantum Presents: Meet the Real Creator–You! (audio), 2005

Dr. Quantum Presents: A User's Guide to Your Universe (audio), 2005

Dr. Quantum's Little Book of Big Ideas, 2006

Dr. Quantum in the Grandfather Paradox, 2007

Dr. Quantum Presents: Do-It-Yourself Time Travel (audio), 2008

Time Loops and Space Twists: How God Created the Universe, 2011

Section 3

Mind and Matter Interaction

Garret Moddel, Ph.D.

Experiments in Psi and New Energy Technologies at the Edges of Physics

Photo by Mulu Moddel

Questions to Ponder

Dr. Moddel was a materialist skeptic about psi phenomenon. What changed his mind?

Retrocausality could explain some psi effects. How so?

Our linear definition of time is wrong. What do experiments indicate about our connection to the future?

Dr. Moddel did an experiment without humans to study the observer effect. What did he find? What is psibotics?

Consilience is a current theme in consciousness studies. Explain.

I was born in Dublin, Ireland, in February of 1954. *So, you're an Aquarius.* My family moved to Southern California when I was four. I had a brogue when I was a child and people would like to come and talk to me, but I've lost it unfortunately. *Tell me about your family's beliefs. Were they Irish Catholic?* No, we weren't really spiritually-oriented at all. I grew up really as a sort of rationalist scientist, as opposed to spiritual.

What led you to study engineering at Stanford? When I was about five years old, I went to my brother's high school Open House. I remember going into his physics classroom, where the instructor had a big Van de Graaff generator and he let me hold a lightbulb up to it. It glowed, which I thought was really fascinating. Later, when we moved on to other parts of the classroom, I went back and turned on the generator again. I've just always been curious and fascinated by how things work. When I went to college, I was interested in a number of different things, but ended up in Electrical Engineering because that's the part of engineering where you're really involved in the process of how something works—if you look at semiconductor devices and technologies like that. That, to me, was an amalgam of basic science and applications. I suppose that I approach life as an adventure and try to use basic principles to invent something that works.

You have patents; you invent things. Yes, I have patents in a number of different things. I've been working (originally in industry and then at the University of Colorado) in energy conversion devices such as solar cells, and other sorts of technologies that are a lot more exotic like trying to harvest zero-point energy. Surprisingly to even me, I have a patent on that. Most recently, my lab has been working on an alternative technology to solar cells that harvests light as waves rather than as photons—-it's a different way to go. It's still going to be a while before it's entirely practical.

There's a lot of mystique about cold fusion, now called low-energy nuclear reactions. I've heard wild stories about the people who have invented them and then men in black suits take them away and destroy their invention. What is happening with this? I am going to say something that probably a lot of people won't like. I've worked quite a bit with low-energy nuclear reactions, with Ph.D. students working on it; one who just recently earned her Ph.D. I worked with a local company that has probably carried out the most rigorous types of low-energy nuclear cold reaction experiments anywhere, looking at a lot of different technologies from a lot of different people and I don't think that there's anything there. I think it's due to the misinterpretation of measurements and overall problems with the experiments. For example, there are two main types of low-energy nuclear reactions. One type-- the Pons-Fleischmann experiment—consists of heavy water, a type of water that's reactive in a nuclear way for harvesting energy, with palladium electrodes in water. Another approach uses gas-based low-energy nuclear reaction, where deuterium gas is absorbed in palladium that presumably produces excess heat, which is the low-energy nuclear reaction. There are a lot of experiments on that, and in fact, when you infuse palladium with deuterium, it does produce excess heat. But it turns out that that excess heat is due to a chemical reaction.

I'm probably saying more technical terms than you're interested in at this point. *I'm interested because this could change the whole planet.* Right, it could if it worked. So, the deuterium takes the place of hydrogen in water in these cells and that turns out to be an exothermic reaction that produces heat. If you just look at it casually, you would think that you just had a nuclear reaction that produced that heat. However, if you look at it more closely, you'll find that it is in fact a chemical reaction that produced that heat and that you can reverse it. If you reverse it, instead of exothermic, it becomes endothermic and it cools down. My student demonstrated this very rigorously; it just is not a nuclear reaction. We also took a look at some other types of cold fusion experiments, and in particular this company that I mentioned looked at it in more detail than my lab did. Over and over, they looked at it and got the same results as my lab did and found that it was in fact due to measurement errors. It turns out that measuring heat is not an easy thing to do; it's very easy to fool yourself.

If you were in charge of a global low-cost high-ecological value energy system, would it be solar? Recently I got a patent for a technology to harvest radiant energy at night. The big problem with solar cells is that the sun doesn't shine at night. Solar cells are very good, very efficient, and quite low cost. The big problem now is energy storage. If you could generate energy at night, as well as in the day, that would solve the storage problem. And now, as it turns out, there is a way to do that. The sun radiates onto the earth at about a kilowatt per square meter, which is used by solar cells. The earth in turn radiates heat at roughly 300 watts per square meter. If you could convert that to electricity, then you'd really have a nice technology.

The problem is, to really convert heat into electricity, you not only need a hot source, but a cold sink as well. It's the difference in temperature between the hot source and the cold sink that ultimately determines how efficient a conversion technology is. If the earth is the hot source, then what is the cold sink going to be? Everything around us is the temperature of the earth, but deep space is cold. If you can absorb from the earth and radiate into deep space, then you have the technology that could generate electricity at night. My patent describes a method for doing that. It is not going to be a very efficient technology, at least not initially, but it's a feasible way to generate electricity at night. I would love to see something like this develop.

Does this method use the same kind of solar panels that we see on roofs? No, because the radiation from the sun is in the visible and near infrared. The radiation from the earth is lower temperature and it's in the mid-infrared; therefore you need a different type of technology. *It's stored in batteries?* It just produces electricity. What you do with it is up to you. *If I wanted to store it because I'm sleeping, I could use batteries?* Yes, absolutely.

After Stanford, you went to graduate school at probably the premier university in the world. Being surrounded by so many hotshots—is it humbling? Is it ego-boosting? What's the psychological feel being a graduate student at Harvard? For two years I was a tutor in a dormitory, called a resident advisor at other universities. I saw a lot of undergraduates and heard their problems and helped experience some of their joys. These students are set up to have strong egos, good intelligence and good connections. For those who fail, it's really sad. It's a harsh environment and those who fail really do feel stupid. It was not my favorite environment because of that. I found, after having both attended an East Coast school and a West Coast school, I think one can generalize that in the East Coast, students were trying to show each other how little they were working and how well they were doing. In the West Coast, students are showing each other how hard they're working and how badly they're doing. *So they don't seem like a snob?* Exactly! You're just like everyone else—it's not about one-upping, it's a camaraderie. *The University of Colorado at Boulder where you teach is more like Stanford, right?* Yes, CU has a West Coast feel, even though we're a thousand miles from the coast.

When have been the most difficult times in your life and how did you cope? A difficult time was when I left graduate school and went against what my advisor advised: "Go see where you can learn from other people and see where you can advance yourself." Instead, I went to a little start-up company in Silicon Valley that was making solar cells, which was an absolute blast! I loved working there, but no one was more of an expert than I was in that particular area. We were all neophytes. I realized—after about three years—that I needed a community of scientists to interact with. I didn't like being in a lab and developing technologies that remained secretive, that only came out as a product. At this point, I realized that I needed to go academic. This was probably one of the most difficult decisions that I had ever faced, to jump ship, and say, "I'm going to take the risk of going academic."

Did you have a family by this time? I know that you have children. No, I was still single at that time. *Was having a family difficult at times? A lot of people say that the most difficult part of a marriage is when the kids are born.* I think that most parents would nod when I say that the most difficult aspects of raising kids these days are their addiction to screens. You hear all sorts of experts talking about how bad these things are, but when you ask them what they've done about their kids they say that they've given up. Their kids are on their iPad and iPhone and playing video games. *How do you monitor this? Do you set limits?* I try. I'm sure not successfully as I'd like to be. *They started with iPads in kindergarten as part of the curriculum.* I'm teaching a new course about new technologies, judging them and looking at the positives and negatives. I was thinking of calling it "Black Mirror." I want to take a look at the screen technology itself and the social implications that come along with what we're doing; they're immense.

I've worked on global youth attitudes and activism for a decade and seen that kids who are educated and have access to screens definitely have a more global and altruistic worldview. Then, I hear stories about children who don't understand how to read body language or facial expression because of their screen use. We've all probably seen kids who are in the same room together, texting one another. Every generation says that the following generation is worse, or that they're lacking something that the previous generation had. They may be lacking some things, but they're no worse than the previous generation. I think that what you said is true, that this new generation is very worldly. Maybe their deficits will be made up with their benefits.

You were on sabbatical when you were exposed to a psi library that got you interested in psi. Tell us how that evolved. This was around 2000. I accidentally came across a physicist's library and was blown away to discover there was a science of psychic phenomenon. Before that, I just assumed that this concept was due to fuzzy thinking and soft minds. As I learned about it, it was absolutely fascinating. I had to get into this and try to understand it. Part of that fascination is that we really don't understand it but it impacts so much of our lives in so many ways.

James Carpenter thinks that we shouldn't say paranormal—we should just say normal. The unconscious mind draws on psi and we're accessing that kind of unconscious information all of the time.* Yes, I completely agree with his perspective, he elucidated it very well in his book *First Sight*. This really is something that underlies our daily existence, once we're aware of it.

How often do you teach your honors course "Edges of Science"? Every two or three years. *What I thought was really amusing was the process to get it approved as a liberal studies course was very complicated and cumbersome and they wanted to see your resume to make sure that you were a real scientist. You felt like your colleagues didn't want to deal with this—"out of sight, out of mind." How much of this is still true?* It's still generally true. I am treated with benign neglect, which is fine. Actually, I came out of the closet in a big way just last month. Our university, along with the world for that matter, is getting into quantum computing, based on quantum entanglement. We had a forum at the university last month in which researchers from across the university described how their research might relate to quantum computing in order to form collaborations.

I gave a talk in which I showed some results from Dean Radin* and Helmut Schmidt, in which intention affected a quantum process. I said that we really needed to be careful here, because it's quite possible that our attention and intention will affect quantum computers and quantum entanglement. I said that publicly and only one person laughed, so I think it was okay. Schmidt was the inventor of the original RNG based upon quantum decay. I discussed Dean Radin's more recent work, in which he asked subjects to use their minds to modify slit visibility in a double-slit experiment. *He's also working with plasma to change it with intention, as discussed in his chapter.*

The key word that underlines all of the psi work is consciousness. How would you define it? With great difficulty. It really does depend upon the discipline that you're in as to how you would define it. I'd like to narrow it to discuss conscious interaction because it's something that's pervasive and affects everything. I've come to that conclusion kicking and screaming. Other people were far more advanced in that conclusion for many years. I thought for a long time that we could understand these psychic phenomena based purely upon physical models, but I've come to the conclusion that really there is a fundamental nature of consciousness in this universe and that maybe at some point we'll be able to understand it in physics terms, but not with the models that we have now.

You've said that to understand consciousness we must include psi phenomena. With non-locality (as when two photons are entangled, they're separated and the spin changed, the other one instantaneously responds), it's very tempting to say there's some kind of quantum information field. My understanding is that you cannot say that. Yes, it's very tempting to use quantum physics models to explain psi phenomena but really, we just don't know. Quantum mechanics does have interactions at a distance, and so does psi. Quantum does involve effects due to not just the past, but also the future and the present and so does psi. It's very tempting to say that one is due to the other. If we had lived at the time of Franz Mesmer (1734 to 1815), when magnetism was being developed, we'd say, "Magnetic fields give interactions at a distance and magnetic fields are mysterious and invisible. So maybe, it's animal magnetism that is the underlying physics that makes this work." In every generation, there is a fashionable new concept that we try to apply to whatever is unknown. Right now, we're doing that with quantum mechanics and psi phenomena. Maybe there's a connection, maybe there's not.

What could be a possible explanation as to why entangled photons instantaneously respond to each other? According to the Copenhagen view of quantum mechanics, that just is the way it is, leading to the expression, "Shut up and calculate." We can't really understand it—it just is. An explanation that I really like is by a physicist who died recently, Olivier Costa de Beauregard, a student of Louis de Broglie. When he learned about quantum entanglement as a student, he came up with a model for it. Essentially, he said that in quantum entanglement, you typically have two particles that emanate from a single decaying particle—they're essentially connected at birth. Whatever you do to one of them is correlated with the properties of the other one instantaneously. That goes against common intuition. What de Beauregard said is that all you need to do is introduce retrocausality. Whatever you do to this particle, goes back in time and affects the genesis of the pairs and therefore affects this other particle. At a distance it looks like the same instance, but in fact it's because it went back in time to make that interaction work. I think it's very valid and valuable. In fact, from psi experiments we know that there is retrocausality. It's not a big leap to say that it works in entanglement scenarios as well. It turns out that de Beauregard wanted to publish this idea and de Broglie didn't like the idea and discouraged it. It wasn't until years later when he was an independent physicist that he did publish it.

In an Israeli study by Dr. Leonard Leibovici, the patients who were prayed for 10 years in their future had better outcomes in the past. He introduced that experiment just to show how you can come up with illogical conclusions by applying statistics blindly to a small study. He actually didn't believe in the result, and in fact, the experiment is a very small experiment that really doesn't prove anything. I fully accept retrocausality, but I don't think that experiment shows it.

What's an experiment that demonstrated to you that it does exist?

Lots of them! Dean Radin* had a number of experiments in which people looked at a screen on which an image that was either disturbing or calming was shown to them. They showed an emotional reaction in advance to the disturbing images as measured by various sorts of instruments. The cleanest experiment that I really like is one that was initially done by James Spottiswoode and Ed May, in which they took the poor subjects and put a horn against them that blasted at a random time, based upon an electronic RNG. They found that the subject who was exposed to a randomly-timed horn had a surge, as you might expect, in their galvanic skin response. This is a very sensitive measure of the emotional state that went haywire after the horn went off. This response began about two seconds prior to the horn going off. It was a very nice, clean experiment. *Doesn't this show that the future bleeds into the present rather than the past influences the present?* If you look from the perspective of that future person or thing, they're looking into the past. It really depends on what perspective you're looking at.

Another principle that applies from quantum mechanics to psi is the observer effect. People are becoming more aware that it's very hard to separate the intention and belief of the experimenter even if they're not in the room. Right, according to quantum mechanics, a particular quantum wave function is in a superposition of all possible states. It can collapse into one final state after it's observed. It becomes particle-like after it collapses. The thought experiment explaining this idea is Schrödinger's cat, in which the cat is in an indeterminate state until it's observed. There are different perspectives on what observation means. Most physicists think that observation means simply that is registered by a detector or anything else. John von Neumann and Eugene Wigner had a different view. They questioned how one can separate the detector from the experiment that it's detecting because it's all part of a system. You can make a bigger system which is detecting what the detector detects, so that the bigger system won't have made an observation until it detects the detector.

You can keep on going, and ask, "Where does this expansion of detection stop?" They said that it only stops at consciousness and it's the conscious observer doing the observation that is necessary to actually collapse the wave function to a well-defined state. This is a minority view in physics and there are some physical theories that I like which don't even need an observer at all. I'm not sure that we actually need an observer, or an observer effect. Other theories that explain quantum phenomena, such as stochastic electrodynamics, and other quantum mechanics interpretations, such as John Cramer's Transactional Interpretation, don't require any sort of an observer [*see Ruth Kastner**]. On the other hand, the observation system has been applied to various physics and psi experiments. One of the first people to do this was Helmut Schmidt. He did a wonderful set of experiments in which he showed that only after something was observed, did it go into a well-defined state. In fact, you could affect what state it went into even days after the process occurred by observing the history.

If the PEAR lab RNG data hadn't been observed, they could change it retroactively. If it had been observed, you couldn't change it. A way that I like to look at this is by asking the question, "Can you win yesterday's lottery, today?" Let's assume that we have a lottery in which the winner is determined by a random process. Today, you don't know who won the lottery. You didn't look to find out. Can you use your intention to affect what happened yesterday and thereby win the lottery today? The answer is yes, we've shown that in various sorts of psi experiments. If you have an RNG that produces numbers that go into a computer register, but nobody looks at them, then you can affect them.

This leads to what is sometimes cruelly known as the grandfather paradox. If you go back in time and kill your grandfather before he produced your parent, then you would no longer exist. I don't like that one, but it's what's called a

Bilking Paradox.[1] You essentially have a future, and then you go back into the past and destroy whatever it was that created that future. This is one of the arguments that's used to show retrocausation can't occur. I think that there's a way around this. Think about this lottery in which you use your intention to win yesterday's lottery. Let's say that you listen to the radio and you hear the first four numbers out of an eight-number lottery. Now you've got knowledge of four numbers and cannot go back and change those numbers because you already know what they are. You are able to go back and change the other four numbers. Your ability to affect the past has to do with your ignorance of it. The more ignorance, the greater the ability.

This is also consistent with the quantum observer effect, in which there's a knowledge perspective. The more you know about something, the less labile that system is. I don't think that the Bilking Paradox is actually a paradox. (The paradox is often stated in terms of bilking, which was originally used to show that retrocausation is logically impossible. If an event A is caused by a subsequent event B, then once A has occurred it should be possible to intervene and block B from occurring. If this blocking could be accomplished, then event A would be bilked out of its cause and could not, in fact, be caused by B. Hence the paradox.) I think that the more you know about something, the less you can affect it. If you already knew that you were born and existed, then you can't go back and change that. You can only go back and change things that you don't know about.

Many studies show that psi skills are high at the beginning when people are interested. When they get bored, the effect drops off—like reading Zener cards. Why are our emotions so important in an intention being efficacious? I wish I knew. This is an area of study and some discussion. Rephrasing your question, is this a psychological effect in which you need the novelty of something happening in order to be able to apply your intention to it? Is it something inherent in the psi process itself that it actually does wane? There are some models of counterbalancing. In other words, if you affect psi in one direction for a while, then nature needs to counterbalance it and produce a little bit of the opposite for a while. I think that this is not resolved.

William Bengston said that he picks students to be his mice healers who are skeptics so there's no ego involved in their healing. A student whose mice died felt embarrassed to be seen with his hands over mice. His feelings got in the way of healing them.* It seems like our emotions do get in the way. In my office I have a Psyleron Mind Lamp, which has a little electronic RNG inside of it that determines which of several colored light-emitting diodes are on. It produces different colors that randomly change from one color to another. The idea with this Mind Lamp is that you use your intention to affect what color is produced. As you can imagine, my kids love playing with it. If I sit there and say, "Go green, go green, go green," nothing happens. On the other hand, if I intend "I want it to be orange" and then go away and do something else and put no emotional investment in it, when I turn around ten minutes later, there is a good chance that it's orange. All I know is that it's not instantaneous, and if I apply intense emotion to it, it doesn't work. It's more of a hands-off intention that does best. *I've found the same in doing clairvoyant work, where neutrality is the key.*

How does the quantum uncertainty principle apply to psi? Once you look closely at something, the other part of it becomes fuzzy. People quote the uncertainty principle as being quantum mechanical phenomenon. In fact, it's a wave phenomenon and you don't need quantum mechanics for that. Let's say that you have a wave of some kind. It can be a light wave, a water wave—it doesn't matter what. If you look very closely at it, you can tell me where that wave is, but you can't tell me the wavelength. The wavelength requires you to look over some distance to see what the periodicity is, and so on. If you look over a long distance in that wave, you'll be able to very precisely tell what the wavelength is, but you won't be able to tell me where it is. Any wave has this uncertainty. You can either tell where it is, or what its wavelength is; that's the same thing that we see in the quantum uncertainty principle. It's two complementary parameters. You can know one precisely, but the other imprecisely. It's a wave phenomenon for any kind of wave. *So, it doesn't really apply to psi directly?* It applies to psi, probably because it's just the nature of physical reality in general. You look closely at something and you see one property. You look at a distance and you can see a different property.

Here's a little question: Physicists are looking for the unified field theory—to unify Quantum Mechanics and the physics that explains the world that we live in. Stephen Hawking was looking for it. Has anybody come close to finding it? This is far out of my domain, but people are looking at string theory as potentially connecting gravitation and quantum mechanics and electromagnetism, and so on. String theory is one of these grand theories that you can do anything with. I'm a little bit afraid of theories like that because if you say something very grand and broad that could be applied many different ways, I don't know that you've said anything testable enough to explain anything at all. I prefer a little theory; just give me a little model that just explains some small phenomenon thoroughly, accurately, and provably, rather than something very grand that's speculative.

ISSSEEM is in your neighborhood, involved with Holos University. What research institutions are doing good science in psi? One extreme is the more experiential or application-oriented organizations. They look at healing and the experience of these phenomena. On the other extreme end is the Parapsychological Association and the Society for Scientific Exploration, which is broader. They look at the basic science and try to understand how we can categorize these phenomena and start to understand the ways they work. My interest is more in the more left-brain approach to parapsychology and psi phenomena. The more experiential organizations satisfy a great human need; however, it's just not my perspective. IONS is a little bit in the middle, more experiential; with a scientific contingent with a number of very good scientists. *Then there are a few universities—like the University of Virginia's Perceptual Studies and the University of East Georgia does some paranormal work with Christine Simmonds-Moore.** Right, and those tend to be more on the scientific side of things, as opposed to some of the more spiritually-oriented universities, which are experiential as opposed to scientific.

You were the President of the Society of Scientific Exploration. A few years ago, yes. It's a wonderful organization and I'm so relieved to no longer be president. *The current president, Bill Bengston,* said he wanted the 2019 conference to focus on what we're moving toward rather than what we're not—what is the post-materialistic paradigm?* The theme was Consilience, "the principle that evidence from unrelated sources, especially science and the humanities, can converge and produce unified conclusions." We want to not just talk about our own scientific endeavors but generalize them and take a look at where this is leading and how it connects with other disciplines. One simple example is if you're a business person and you find several synchronicities in your life, might that lead you to a conclusion that can help your business? I believe that the answer is yes, it can actually be quite useful. We are trying to bring up this issue, not just to the psi community but to the general science and even the business community.

Why would a businessperson care about psi research? There are actually a number of businesses that use psi advisors of one sort or another. In some cases, it's something like astrology. In other cases, there are various sorts of remote viewing and other techniques applied to making business decisions. I think this extends a lot farther, to human interactions in a work environment. It extends to following our hunches. Taking this out of the realm of urban legend and putting it onto a firm scientific foundation will allow people with various disciplines to say, "I actually have observed these effects; they are useful and I will continue to work with them."

In terms of applied psi, you and your students did some work with Associate Remote Viewing to predict the stock market. You did it with symbols, which the viewer didn't even know what they meant. Seven out of seven times, you got it right. This was taught to us by Paul Smith,* who's one of the Star Gate trained remote viewers. He used a technique that was originally suggested by Stephan Schwartz,* called ARV. Let's say that you have a person who is tasking people to do some remote viewing and draw pictures of an image. When you task that in your mind, you associate two images with a future event. For example, a picture of an orange might be associated with the Dow Jones Industrial Average going up tomorrow while a picture of a pencil might be associated with it going down.

You don't say anything to the people that you're tasking, other than ask them to draw a picture of what you will show them at the end of tomorrow. The judges take a look at the images and ask if they're more orange-like or more pencil-like. That's actually a fairly easy judgment to make because the two options are so distinct. In this class, based upon whether the majority saw it was a pencil or an orange, we decided that the stock market was either going to go up or down the following day and we invested accordingly using day trading. Then, at the end of the day, we sold, and based upon whether the market did go up or did go down, we showed the people the image associated with the actual outcome to close the loop.

We did these seven times, and seven times out of seven, we got it right, and made a fair bit of money. The problem is that as with most psi phenomenon, there's beginners' luck. It works for a while and then there's a decline effect. Whether that decline effect is psychological or something deeper, I don't know. A number of people have tried to find techniques to avoid it, with very little success. You can beat the odds, you can do better than 50/50, but you can't get anywhere close to 100%.

The Star Gate remote viewers did that kind of work with silver trading and were successful. I don't know of anybody who's done it over time. It may be the same concept as the colored lamp—if you put too much energy on it, it blocks the effect. Maybe, maybe. A paper published in the *Journal of Scientific Exploration* by Debra Katz took a look at predicting markets over time. She found that over time there was this decline effect and that you could not beat the odds. *Why not just have a team of new people cycling in all of the time?* Maybe that's the way to do it. You may need to not only have a new team, but you may need to replace yourself, too, because as the experimenter, you'll want to be completely hands-off.

Speaking of hands-off, you wanted to see if you could have that kind of effect without any human beings, by having an RNG connected to a computer. For years, I've been trying to think of a way to get rid of any living beings to design an experiment that involves just machines to demonstrate psi. Finally, I thought of a way to do it. Essentially, it would be an inanimate replica of that Spottiswoode and May experiment. I would have the controller RNG that could shut off a subject RNG while it was spewing out random bits, which would be recorded by a computer—zeroes and ones. The question was: Could the subject RNG anticipate its own demise by putting out a non-random set of numbers before it was being shut off? A graduate student in my lab worked hard on this. James Zhu, a graduate student in my lab, worked hard to develop the software, and Adam Curry, who at the time worked at Psyleron (the company that makes the Mind Lamp), built the physical apparatus for this. Finally, we set it up in my lab. James and I were so excited; we were tickled to finally be able to do this. We ran a few hundred runs, and low and behold, about one second before the subject RNG was being shut off, it started producing more zeroes than ones. We did this again and again and ended up with a few thousand samples in which the odds against chance of its being a random effect were millions to one.

We were terribly excited because we finally got a way to get rid of these sentient sacks of saltwater and were able to just use the electronic RNGs. Then, we went about testing different bit rates and changed the scheme a little bit. It didn't work the way that it did when we originally did it. After about two weeks, we went back and replicated the original experiment with the same software and hardware and got no effect whatsoever. At that point I realized that this was not the machine producing a precognitive effect—it was us––our intention. We had imbued the machine with our intention. At that point it finally dawned on me: Garret, stop trying to create such effect with just machines. There's something about consciousness that's a lot more profound here. I don't want to go to the spiritual realm, but I'm sort of being nudged that way by our experiments.

What about your experiment where the Zener cards were drawn by a human remote viewer and a machine? In a RV experiment a tasker asks an RVer to draw an image of something that they will observe later or it's associated with a number or a name or something the tasker gives the viewer. *Or just an envelope.* Right, or just an envelope with instructions to draw what's inside of the envelope. A student of mine at the time, Erik Maddocks, set up an experiment that I'd been wanting to do for a long time. We had an envelope which contained one of five Zener cards. We know the Zener cards from having watched *Ghostbusters*: a wiggly line, circle, cross, square, and star. The tasker tasked the RVer to draw what was in this envelope. The viewer didn't even know that it was Zener cards.

Unbeknownst to the RVer, in the room we also had an RNG that was spewing out bits which were being recorded by the computer. Then, after the remote viewing was done, Erik had a program that took those bits and created a raster scan image based upon the bits in the computer. Ones consisted of black dots and zeroes consisted of spaces. We had a series of dots creating an image based on the output of the computer, fed by this electronic RNG. Then, the judges examined the image to determine which of the five Zener cards the RVer had drawn. They had two different images to judge—one was what was drawn by a person and another one drawn by the computer. It turned out that in both cases, the RVer was able to get a statistically significant result in getting the right image. But actually the computer-drawn images were more accurate than what the RVer had drawn. We call it "machine-mediated remote viewing," where you get the machine to draw the remote images. This was an initial set of trial experiments. I would love to repeat it and refine this. It would be good to improve this by having the computer do grayscale and various sorts of image enhancement. We just did a very simple grid.

Your students did an experiment with grass seedling growth, affecting it with anger. This was a student of mine in the "Edges of Science" class. Each student had to carry out a research-quality study. For his research, he got 100 little seedlings, each of which was put in its own pot, divided into four groups. All of them were kept together except for 15 minutes a week, when one-quarter was taken out. One group was the control; they weren't subjected to anything. Another fourth was subjected to torture of other plants. He took a plant, crumbled it up, and put it down the garbage disposal. Another group was subjected to his swearing at it for 15 minutes. This guy could really swear. The other group was both subjected to the plant cutting and the swearing. He called it "mean speak."

After a few weeks, he measured the germination rate and the height of these four different groups. He got a huge difference between the three groups that were berated in one way or another, versus the control group. He got odds against chance—I think the p-value was something like 0.0007. Odds against change were 100,000 to one and he was seeing an effect. *So, the control group grew faster?* Yes. *The plants that were cursed at and exposed to torture—they were the worst?* All three groups that were mistreated had similar outcomes. *What about a fifth group of plants told, "You're so green and beautiful, I love you." You would see if they grew any faster than the control.* In my last class, one group actually did do that. They saw an effect, but we can't make much of it because they didn't get enough plants to get statistical significance.

You did an experiment where glass reflects light and with your intention, you decided whether it was reflected or transmitted. The idea was that when light goes through a window or a glass slide, 4% of the light is reflected at the front interface and 4% is reflected from the back interface. If you take a look at this from the photon point of view, you don't know in advance which photon is going to be one that will make it through; which of the 92% that will go through and which of the photons that are part of the 8% that will be reflected. This is a random process, and so with intention we should be able to affect that. We set up the experiment and measured the reflected beam and then the transmitted beam and then applied intention. We did find that we could get a significant effect in shifting the amount that was transmitted versus the amount that was reflected.

Did you do it as a group? If so, did you find that group intention was more powerful than individual intention? My students took one individual at a time, sitting in front of the computer and the reflection apparatus. We have actually done group experiments with a different intention. I invited a very prominent skeptic to my class—Victor Stenger has written several books on skepticism and how God doesn't exist. To his credit, he participated in some of the experiments. We found that whenever he was a part of the group that participated in the experiment, we either got a "null" result, or we got a "psi missing" result, meaning the opposite direction of intention. When he wasn't in the room participating, then the psi experiments worked as expected. That really does show that humans affect the experiment.

What about psibotics as discussed in a video of your presentation to the Society for Scientific Exploration? That has a lot of implications for health and other applications. How does it work? The basic idea behind psibotics is to use psi, to use your intention, to program a machine that has some sort of a random process in it and thereby use that machine to provide a desirable function. We've already described two psibotic machines--the one where we had two RNGs, one of which shut off another, and that was used to predict the future. That is a psibotic machine because we were using our intention to get the second RNG to respond to the first RNG. If we had set it up to look at what was going to happen a few seconds hence, we would have been able to predict the near-term future using that machine. The other psibotic machine was when we used an electronic RNG with remote viewing, where the machine mediated the remote viewing by the electronic RNG through the image rather than having a human do it. Those are two examples of psibotic systems.

I proposed this whole notion of psibotics because it would be wonderful if we could actually use this. I'm not the first one to do this; other people have used little robots controlled by electronic RNGs to perform various functions. The question is, can we do this reliably enough to provide some useful function due to the two big problems with psi experiments—the decline effect and the experimenter effect. That is, things tend to decline in time although they may build back up. Also, it's very hard to know who's controlling what, because everything has an influence on the machine. What can we do, both psychologically and physically, to our systems to enable them to work in a more reliable fashion? I don't know the answer. I've got a few ideas, but we really need to work them out.

When paraplegics are set up with electrodes in their brain, they can think "lift my leg muscle," but that just requires a certain brain wave?

Right. That's not psibotics at all; that's just the machine picking up some sort of brain wave which is then translated into motion. *You have to have a device to pick it up, whereas with psibotics, you don't need that intermediary device?* Correct, you're using intention to affect a random process.

So ideally, someone who is a paraplegic would be outfitted with something where they think "walk," and then they walk? I wouldn't want to use a psibotic system for that, simply because it's not reliable enough. Maybe 60% of the time, they would move forward and 40% of the time they would move backward, so we really need a more reliable system.

Speaking of the power of thought, why is placebo getting more effective? If you take a look at drug trials for, say, antidepressants, there is evidence that when you run a new drug against a placebo, it does less well than if you run the new drug against some old drug. I believe that drug trials are often done these days not against placebos, but against other drugs, because the placebo has become so damn effective. Is that a psi effect? I don't know. It certainly could be. It could be that the knowledge field, if you want to speak loosely, of how to use antidepressants to affect depression is out there. You don't necessarily need the actual drug itself.

Bill Bengston found that the cradles where the healed mice were kept became resonant with the healing. Thought forms do create some kind of change over time?* It looks like that, yes. As to what the actual process is, nobody really knows. Bengston is unusual in being able to find this very reliably. Is it actually some sort of a resonance field in that environment, or is it somehow in the minds of some of the researchers? Or maybe even the readers of that research later? Who knows? *Bengston said that the water in places like Lourdes, that people think are sacred healing places, has a different composition. Belief does have an effect. That would be an interesting project—to go around and measure the water in sacred sites, or "holy" water in churches.*

What's interesting to me is that Radin's research, your research, and others have effect sizes that are thousands of times beyond chance, yet they're still dismissed. So, is it human nature that we oppose a new paradigm, or is it something about psi that makes people so reactive? Certainly it is human nature to not want to change our minds once we get a particular worldview. I taught a course called "Science Court," in which students chose one of two sides of an argument to debate. We used courtroom procedures, along with a judge, jury, cross-examination, and so on, to look at issues that were scientific questions of public interest.

It turned out that the science court was a failure, in the sense that the people who were better at arguing won, rather than the side that was more correct. I tried changing this by having people change sides partway through the argument and using some sort of mediation technique to get a consensus. I found that once somebody had argued for a particular side, they would not let go of their original opinion. Everything else that they observed and saw was formed around confirmation bias. It's really hard to get people to change their minds. We, in the last few hundred years, are living in a materialistic era in which we believe that everything that is, is material and that anything that is nonmaterial, such as consciousness, is just soft thinking. It's very hard to get away from that.

In support of people who are skeptical about these experiments and these results—we don't have a good model to support it. There are a lot of ideas about consciousness being fundamental in a multidimensional universe with entanglement of objects over large distances and times, but none of them are useful at this point. We need clear and precise models that can be tested. According to Karl Popper, if a theory can't be tested and falsified if incorrect then it has no value. If we had a good model that we could verify, it would be easier for people to jump in and say, "Yes, I accept this." But, as it is right now, self-respecting mainstream scientists, and psychologists in particular, refuse to accept psi publicly—privately, they may. Something has got to happen differently for people to shift their view. Maybe somebody starts a company and makes a lot of money using psi phenomenon––that tends to shift people's view. I'd like to see us develop models that make a lot of sense and apply those models for psi phenomena.

Do you think that it would be possible to make a mathematical model for psi? A testable model will probably be expressed mathematically. The main point, however, is that it must be sufficiently quantitative that it can be tested and shown to fit the data or not. Simply having conceptual models is not enough. *String theory is a mathematical model, but there are no experiments that prove it, right?* As far as I know, yes. To your point, I think we need the model to make some sort of physical sense to us. It's not just pure mathematics, but it has to have something that we consider an explanation.

Do you find that you have more synchronicity and intuitive flashes because you're thinking about these kinds of things? Before I uncovered the literature on psi phenomena, I thought that this was all rubbish. Since I've come to read about it and perform experiments, yes, I am seeing these effects around me fairly often. I think that once one becomes sensitive to it, it's no longer an abstract phenomenon but an everyday real thing. *What's an example?* I was taking a hike with my brother in Ireland, and in the middle of hiking up a mountain in the Burren, he said, "Garret, call home." I asked him why, and he said, "I don't know, just call home." So, I called home, and found that my department chair and several other people had desperately been trying to reach me, as there was a new institute being set up and they wanted to know whether I would be the director. They had tried everything and didn't know how to reach me. The wife of the Chair had actually broken into my house to try to find some phone numbers where I could be reached. My brother has never said something like that before that, or after that. What was that due to? I don't know.

What's fun and renewing for you? I like drumming, hiking, and music of various sorts. But one of the things that I enjoy most is reading and thinking about science and psi phenomenon. My bedtime reading right now is a book on quantum mechanics. This really is fun for me! *Many of the scientists I've interviewed say the same and also mention nature.*

We have around 11 years to turn back the tide of climate change; we see autocrats being elected around the world and increasing inequality. Are you an optimist or a pessimist in the face of this? Right now, I'm a pessimist. I see the rise around the world of totalitarianism, and the intentional confusion of fact and fiction and find that very disturbing. Unfortunately, I think it's going to have to get worse before it gets better. This really is a worldwide phenomenon right now. It reminds me unfortunately of the times that preceded some of our world wars, where truth could not be separated from fiction. *What makes me hopeful, is that young people—like the Millennial women who are elected to Congress—are not afraid. That's very encouraging.*

Publications

Quantum Engineering Laboratory (mainstream work): http://ecee.colorado.edu/~moddel/QEL/index.html

Psibotics Lab: http://psiphen.colorado.edu

Complete list of publications (see CV): https://www.colorado.edu/ecee/garret-moddel

https://scholar.google.com/citations?hl=en&user=VOWsHaQAAAAJ&view_op=list_works&sortby=pubdate

Psi publications: http://psiphen.colorado.edu/Publications.htm

Endnote

1 http://adsabs.harvard.edu/abs/2011AIPC.1408..235D

Dean Radin, Ph.D.

Evidence for Psi Phenomena

Photo by Jordan Engle, Lens&Soul

Questions to Ponder

What motivates Dean Radin to work long hours in a controversial field?

Non-locality is a possible explanation for the physics of psi phenomenon. Explain.

What lab experiments demonstrate the influence of mind/intention on matter?

What has Dr. Radin learned from esoteric traditions East and West?

I was born near the Bronx in February 1952. *Does it mean anything to you that you are a Pisces?* I know what it means astrologically and there does seem to be an overlap with my personality and the traditional meaning of a Pisces. But I can't say that I know enough about astrology to know much more than that. My Myers-Briggs is INFJ.

What led you to your interest in science? It must be inherent because I've always been curious about everything, so like a lot of curious kids I had science kits and read *Scientific American*. I also spent a lot of time reading fairytales and science fiction, much more so than anything related to sports, which I basically never did, partially because I got rheumatic fever when I was 12 and was put on restricted physical activity all the way through college. At the time, they really didn't know how to treat it other than through restricted physical activity and taking penicillin three times a day for the rest of your life. *That probably counts as one of the most difficult challenges that you faced?* Yes, but as a kid you don't know any better since you have nothing to compare it against. I was never all that physically active anyway, so I didn't regard it as much of a loss. The closest I got to physical activity was playing the violin, which was my primary activity for many years. *You worked professionally as a violinist for about five years?* Yes. *People say that music and math ability are left brain, they correlate in their abilities. Do you think they fed each other, your interest in science and in engineering and music?* Probably, yes. I went to a music camp in the summers and at least every other person there was as much of a math whiz as they were a music prodigy.

Why did you decide in college to go into engineering rather than music? Partially because I found playing four hours a day physically exhausting. Fortunately, I was also good in math and science and I soon found that you could make a living with your mind without having to be a musician, which essentially requires you to be an athlete. *Do you still play the violin for fun?* No, because in order to do that I'd have to practice about an hour a day to keep in shape, and I don't have the time.

You grew up in an agnostic Jewish family. It sounds like you grew up in a family that wasn't particularly interested in religion or *spiritual discussions.* I don't recall any discussions at all about these topics. The only exception was that occasionally my dad would say something about religion being stupid, in the sense of all the suffering and violence caused by minor differences in religious ideologies. [Like *Russell Targ's* father.*]

How did you get interested in parapsychological phenomenon? Probably for the same reason that so many people have fallen in love with Harry Potter. It sparks the imagination; it fuels some of the wish fulfillment that the kids have. And because I also had a scientific bent, I was curious about why these stories were so pervasive in literature. I thought that maybe the lore and the mythology were based on something more than a simple fantasy or superstition. I had a lot of time to read and I soon discovered the literature of parapsychology and was very pleased to learn that there were systematic ways that you could directly test whether or not these stories had any basis in truth. That's where it really caught my attention.

You majored in electrical engineering; I assume that as an undergraduate there wasn't much opportunity to do research into the paranormal. True, but we were able to do projects of personal interest. I can't say I knew what I was doing at that point, other than what I had read in books, but I was dabbling with little experiments even before high school. *Like the Zener cards used by J.B. Rhine to test for ESP (available online to print at no cost)?* Yes, or a simple precognition test where you might have a card with an arrow pinned on it, you'd spin the arrow and make a guess where it would land and then you'd spin the arrow. So it was a precognition test. I have a piece of paper from when I was about 12 years old where I was graphing how well I did on that test over time. A scientist would look at that graph paper and say, "This kid is going to be a scientist," because when you start graphing things and get interested in collecting and analyzing data, that's what scientists do (at least that's what empiricists do).

When you went on to graduate school, why did you shift from engineering to psychology? When I was in the master's program, I was pretty sure I did not want to be an electrical engineer. It wasn't because the topic wasn't interesting––I still use a lot of those skills––but because the gender imbalance was so extreme that it didn't feel right. It felt wrong to have 99% of the people engaged in a discipline only be men. The other reason was that my interests had evolved toward artificial intelligence and at the time there was no AI curriculum. I could have gone into development of specialized AI hardware or the software or cognitive side of AI. Only two professors at my university were interested in AI at the time, and I decided to go with the one who was interested in the cognitive side. He was in the Department of Educational Psychology, which was far from what I had been studying. But I decided to switch over to work with him and it took two years to catch up with all of the psychological research and statistics courses that I hadn't taken. In retrospect I am very glad I took those courses so I was able to draw from both a scientific and technical background as well as a psychological one.

An Oxford University study predicted that almost half of the jobs will vanish as AI takes them over in the next 20 years.[1] I think it's true. Robots are getting much better than they were in the past largely because of computing technology, so a lot of jobs can and probably will be automated. If you have a robot doing somebody's job, even something like a waiter, those jobs are definitely going to go away and it's a big problem. *If someone hired you as a consultant, what would you suggest that we start doing now to provide a livelihood for all those displaced workers?* The jobs that will not be replaced, at least in the short-term, will be ones that require creative problem solving. So the solution to this problem is basically a political one; for example, in the US the corporate board of directors and the CEO are bound by law to make a profit. It's not only that people are naturally greedy but that greed is built into our legal system. This is going to require a societal decision to change the law, which is never fast nor easy.

To get back to the spirituality and religious beliefs, you were raised in kind of a tabula rasa where you could shape your own thinking. I know that you are a meditator, how did your spiritual journey progress from your roots? I fit into the spiritual but not religious category, mainly in the sense that the more I learned about the way that the physical world works, the more I realized how much remains unknown. You can look at the outer limits of what we are taught in schools, especially from a scientific or an engineering perspective. We are able to make fancy tools like computers. But science and human history is like a flash in the pan, and while it seems we've come pretty far, in fact we've just begun to scratch the surface of what we actually know. My sense is that there's much more, not only about understanding the physical world, but understanding the nature of consciousness as well.

About 5% of the universe is matter that we have some understanding about and the rest is dark energy and dark matter, which we don't understand. Do you have any kind of sense of what it might be? If our ideas about particles and the whole structure of physics today are even off just slightly, then our cosmological models are completely off too. Alternative ideas about dark energy and dark matter basically say that those concepts are illusions based on mistakes in our current theories. We just don't know yet.

Is it accurate to say that quarks are the basic building blocks and there's potentiality whether something becomes a particle or a wave? If you look at the interpretations of what quantum mechanics means, the ontological basis of it, it's all over the map. It's certainly correct in the sense of all the experimental tests that we've been able to come up with to verify the mathematics, but in terms of what it means as an ontological reality, we don't know. Maybe I'm more willing than most academics who are obliged to defend their ideas to say that I'm comfortable with the ambiguity. I feel it's way too premature to insist that we already understand the nature of reality. This allows me to work in a field as controversial as parapsychology because, based on the experimental data, something is there. But in terms of what it is and how it works, we don't know.

The most interesting implication to me of quantum mechanics and field theory is non-locality, where entangled particles influence each other from a distance, as you write about in Entangled Minds. Some people say psi phenomena like telepathy or clairvoyance are impossible because they violate what we know about physics. To this I would say, "What do you know about physics?" Generally, what they're thinking of is the Newtonian version of 17th century physics. *Quantum Enigma: Physics Encounters Consciousness* (2011) was written by Bruce Rosenblum and Fred Kuttner, physics professors at UC Santa Cruz. Their book is about the relationship between consciousness and its role in physics, mainly revolving around the notion of the quantum measurement problem. I asked them before they started writing that book, "What do you think about the data in parapsychology?" They both had the usual response, "It doesn't look very good." After they finished their book, I asked Bruce the same question. He admitted that he was more impressed with the data because, as I put it, "Psychic phenomena look a lot like the experiential version of quantum non-locality." They involve events that definitely happen but aren't tightly constrained by ordinary space or time.

What do you think is the reason that entangled particles or people are able to communicate? Not communicate, but correlate. Classical correlation occurs when two things move together but they're not connected in any way. An analogy is that you could have two kids on a swing that are both pushed at the same time, so they're going to be moving back and forth––they are correlated. It might look like they are connected, but they aren't. They just had a common cause. The difference with entanglement is that if two people are swinging on a swing and one decides that he is going to suddenly twist to the right, even though the other one is not connected in any ordinary way, she will start twisting too. We understand how this works in a mathematical sense, but so far no one understands the ontological implications of these strange but very real "interconnections." Parapsychological phenomena strongly suggest that there are some kinds of holistic interconnections between everything.

Quantum mechanics would agree with that, right? Yes, but there's a problem in that we don't actually understand what's going on underneath the behavior of the effects that we can see. *Is it fair to say that most scientists accept that there is a universal information field, the quantum field?* Many would say that, but not all, mainly because there are still many physicists who are much more in alignment with Einstein's belief that we live in a localized reality.

Many visionary scientists regard consciousness as an underlying concept, but they say there is no real theory of consciousness. There is a materialistic view and a non-materialistic view of consciousness; how would you define consciousness? For most of my working career, I didn't realize that I was a materialist because I didn't spend much time thinking about the philosophy of science, or the assumptions underlying our ways of thinking about reality. But after reading enough about Eastern philosophy, writing my third book, *Supernormal*, and then reading much more about the history of esotericism, I guess I am an idealist. *In the Platonic sense?* Yes, in the sense that consciousness is fundamental; it's a sort of substance that comes before space and time. I'm not throwing away materialism, but rather expanding it in a new way, adding a new assumption to it. The reason why I think this is an interesting approach is because it solves in one stroke a lot of the otherwise anomalous effects that we see in psi research.

For example, clairvoyance is the experience of gaining information about something anywhere in space or time if you can focus on it properly. From a materialistic point of view, the only way that we can do that is through signal-passing, like with electromagnetic fields. If you try to use conventional fields and forces to model how something like clairvoyance can exist, then you have to imagine that your mind can leap out through space to a specific location, grab information in some way, and then bring that information back into your head. It gets tough very quickly to try to figure out how something like that could work.

The other approach takes an idealistic perspective, where consciousness exists before the notions of space and time. From that perspective, your awareness is not "in" space and time and we already know how to manipulate our awareness through focused attention and intention. So we finesse the materialistic idea of having to use signals, electromagnetic or otherwise, because some part of your awareness is already extended through space and time. All this requires revisioning what perception, cognition, and awareness are all about. From an academic point of view, this is of course an extremely radical position. Only philosophers who are interested in idealism like this idea, but that's the direction I think where the data are pointing.

I was trained in a year-long program to do clairvoyant kinds of work. A recent example is as I was doing chakra balancing with a young man, I told him, "Your mother (who I did not know) is really worried. You should call her." Two minutes later he told me, "My mother texted that she is worried about her back problem." How could I have known that? Is it somewhere between clairvoyance or precognition? It's really difficult to create a physics-type model that would explain that, but if you start from the point of view that your awareness is already spread out everywhere, then you are aware of that in the same way that you would be aware of hearing your name mentioned at a party. It's simply that a portion of your awareness was focused and that information bubbled up to your conscious awareness. *I was entangled with my readee but had no entanglement with his mother so he was a vessel to tune in onto her emotional need?* In a reality that is fully interconnected, nothing is truly separate.

So it wasn't necessary to have met the mother. You just needed your attention to be focused in the right direction.

In terms of your own personal spiritual practice, what do you do to keep centered, especially when you have to be so courageous to put up with so much opposition to the new paradigm? It's like putting Galileo in house arrest; it really drives people crazy to have their paradigms assaulted like you do. I meditate sometimes very diligently and then I'll drop it for a while and then I'll pick it up again. I've been doing that for 40 years. *What form of meditation has worked for you?* After trying many different methods, I've settled on Vipassana. It's more getting into a state of turning off the brain's default network (in neuroscience terms), where thoughts just stop. *Do you use a mantra or any kind of focal point or you just set your intention, "I am going to go into bliss?"* I don't experience it as a bliss. It's like being in a nothing state, except that awareness is still present [*similar to Peter Russell**]. The method involves becoming physically calm and then calmer and calmer. It pushes the idea of physical and then mental relaxation to extremes, where you think you can't possibly go any farther, and then you just keep going.

As far as working goes, I don't feel that I'm working all that hard. What I'm doing is following a passion. I work 70 or 80 hours a week, driven almost entirely by curiosity. Of course, I've learned that if you start challenging other people's ideas, especially in the academic world, you're going to attract a lot of criticism. If you can't stand that kind of heat, then you just stop doing it. *You went from a university to IONS.* Yes, I work at IONS and it's very nice to be among other people who have similar interests. Elsewhere there's a lot of gnashing of teeth that goes on, and for the most part, I just ignore it unless a critique shows up in a published article somewhere. Then I have to respond. But that's pretty rare.

Do you think there's a loving God or does that not really matter if one is doing their best? As a personal entity, especially one that is concerned about me as an individual, no, that concept doesn't make any sense to me, nor do I see any evidence that there's any "super being" like that. On the other hand, if you take the notion of idealism seriously, then the idea that there can be embodiments of consciousness, or localized bodies of unconsciousness in many forms, becomes much more plausible. From that perspective the Earth itself might be a conscious entity, literally self-aware. So could the sun, so could the galaxy, so could the entire universe and then all the way down back in the other direction to maybe electrons have a certain social life. So, I think of God as more like a principle, maybe there's something at the top of that hierarchy, maybe not, I don't know. Again, this is where being comfortable with ambiguity is part of my personality.

When have you experienced the most challenge, the most difficulty, and how did you climb out of it? When people in authority think what you are doing is bullshit and they are highly motivated to do everything they can to make you go away. This involves both the heads of laboratories and heads of universities. I've learned the hard way that administrators fear embarrassment more than they fear death. If you are a president of a laboratory or a provost of a university and you are trying to build a reputation and somebody is getting way too much attention because of the controversial work that they are doing, you will do everything in your ability to make it go away. That's not much fun being on the receiving end. The only thing you can do is go somewhere else, so that's what I've done when I've been in those situations. Fortunately, every time that has happened, it's resulted in a major advancement in my career. I've been at IONS now for 18 years or so. It took a long time to find a place that was both scientific and supportive. Unfortunately, it's also one of a very small handful of places in the world where that combination is even possible.

Where are other research centers? The Rhine Research Center, the legacy of J.B. Rhine, supports a few people who are doing interesting work. [*See John Kruth.**] The University of Virginia's Division of Personality Studies is the legacy of Ian Stevenson. They have a good group doing both experimental and case study research. *About children's past-life memories? I wish they had agreed to be in the book.* Yes, and NDEs. In the UK, there are a dozen universities with faculty who have interests in this topic. The University of Northampton and the University of Edinburgh come to mind. In the US, academic centers with at least one faculty member openly interested in parapsychology, besides the University of Virginia, are West Georgia University [*Christine Simmonds-Moore**], the University of Arizona [*Gary Schwartz**], and the University of Colorado at Boulder [*Garret Moddel**].

Let's look at some of the different phenomena that you've studied like RNGs, that are so interesting because you can't say it's placebo when you get anomalous results. It's true for all of the other phenomena as well. You do a Ganzfeld telepathy experiment or look at correlations between brains at a distance and by the experimental protocol you can eliminate expectation effects. An RNG seems like it's purely mechanistic, yet it also seems to be reflective of something like a collective unconscious. All of these experiments have been developed over many decades, so if anybody found a loophole in the design, it was eventually closed. Now the primary classes of experiments are as tight a methodology as it is possible to devise.

As an academic, what I thought was funny about the RNGs was that at academic conferences, there is no coherence, unlike spiritual conferences. They hardly ever do. Academia is largely driven by ideology, where people analyze things all the time, comparing your ideas against somebody else's ideas and thinking hyper-critically about it. That kind of behavior does not engender coherence, whereas a large-scale meditation does. High mental coherence, from a parapsychological perspective, appears to give rise to physical coherence.

I read that the biggest collective consciousness effect was the World Trade Center bombing and it started three or four hours before the first plane hit. It was a statistically significant effect, but not the biggest deviation in the RNGs that were obtained in the Global Consciousness Project [GCP, *see Roger Nelson**]. It was an unusual moment, like most terrorist attacks, where there is a moment of collective shock that reverberates around the world, but then the collective attention falls apart very quickly because people are drawn back into their own problems. They start thinking in a million different directions, so there's very high mental coherence for a short period of time, and then it dissipates. Similarly, something like World Cup Soccer, which might have a billion people watching, might seem to be an interesting event to follow, except that there are half a billion minds pointing this way and half a billion pointing the other way so any coherence created cancels each other out. By contrast, there's the opening ceremony of the Olympics, where a billion people are focused on the same event.

The formal database for the GCP involved 500 events; that experiment ended a couple of years ago. While there are variations in the results associated with each individual event, the overall effect accumulates to seven sigma (a measure of statistical deviation), associated with odds against chance of about three trillion to one. It's very clear that cumulatively you can see that the mind-matter relationship is a real thing, but in any given event it is very difficult to predict if it will show a big effect or not. The one characteristic that seems to produce larger effects is events that produce very strong periods of collective attention.

When you were at the University of Nevada in Las Vegas, did you see people who win at slot machines? It's probably not about influencing the machine per se. Typically they get the sense of, "I need to go play that particular machine that I know is waiting for me there," and then they go to it and win. Maybe they are getting a hint from their future when they win it. I don't know. *What about remote viewing? Were you able to make accurate drawings?* I usually do best in experiments involving mind-matter interactions. I have done remote viewing, mediumship and clairvoyance as a result of experiments either run by me or someone else.

What else are you studying at IONS besides RNGs? Currently in the laboratory we are using electrical plasma as a target. I am also doing experiments involving photons [*like John Kruth* and Patrizio Tressoldi**]. We are doing experiments with a new form of RNG that we've developed, which is not creating random bits, but rather recording the underlying quantum noise directly. The way you normally get random bits out of an electronic circuit is through digitizing and processing the noise produced in semiconductors. You sample the noise and chop it to turn it into a digital form. We do see effects with that type of generator, but if you see a significant deviation from chance, it becomes very difficult to figure out what caused that deviation. These devices are designed in such a way that they are impervious to environmental effects. That's the whole point for using them so no ordinary influence will cause them to misbehave. But that also means you can't reverse-engineer the results very easily. So we created a new type of RNG where we record the noise before it is turned into digital bits. Then we are better able to analytically figure out what happened to the noise to cause it to deviate from chance.

I wrote about this in *Real Magic*. We had a system of 32 RNGs running during the presidential election in 2016. We saw a big effect right at the time that the election was called. Analytically what you see amounts to a wrinkle in space and time. That's because you can look at the data both as a space-separated phenomenon, because there are 32 separate generators, and also as a time-separated phenomenon, because the data were recorded in time. What you see is that the data were merrily going along and suddenly the spatial connectedness spiked when the election was called. Something very similar happened in a temporal sense. So when you take both of those effects together, it's very much as though there was a warp in space-time when a few hundred million people gasped in horror or delight at the results of the election.

So the warp allows bypassing the normal process? No, it's more like space itself is warped as a result of millions of minds focusing on the same thing at the same time. In space-time, this concept suggests that the fabric of reality somehow goes whoop, and so from that perspective the RNGs are actually not doing anything unusual. They are just going along for the ride and surfing a ripple or warp in space-time, very similar to Einstein's topological view of gravity. That is, the moon is not being pulled toward the Earth because it's orbiting a very straight line. But space itself has been warped, so it's following the shortest route it can in this little pocket of distortion in space and time. Something like that may be a way of understanding why any kind of intentional influence happens.

This may not work as well for things like distant healing because in that case you could potentially influence the other person's body, but you could also do the same thing through telepathy. So distant healing might be due to a "distant placebo effect." *That's why they have studies where one of the control groups doesn't know they are being prayed for or whatever?* Right. I write about this in *Real Magic* because the moment you pray for somebody else for any reason and they don't know about or haven't given you their permission, that's the definition of black magic. *The drawings of the remote viewing done during Star Gate, which you were involved with, are so accurate and so informative. But then that whole program was dismissed as not being effective. I wonder if the government still does it secretly because it is effective?* I suspect there is no official program. But I also suspect that there are many branches of government and military intelligence and businesses that have hired consultants to do what was once formalized in the Star Gate Program.

What was your role in that Star Gate Program? I was a visiting scientist for a year and then I was offered a permanent position but decided not to take it. Part of what we did was threat assessment, which meant that we would read an article about psychic phenomena that had been translated from a Chinese or Russian source and would try to evaluate the credibility of those reports. On a few occasions we spoke to scientists who defected to the West who claimed to be doing research in psychic phenomena. I also did a couple of experiments in the lab to test ideas about the nature of precognition.

I used to get these snapshots of events that would happen, like when I went for a job interview, I saw something that I'd seen in a mental snapshot and I knew I would get that job, but that I would only be there two years. At the end of two years of teaching there, I thought, "I am staying. I don't have any intention of quitting." But then I went to Europe in the summer, decided I wanted to go to grad school in Religious Studies and quit. There's a bit of a mixture there between a precognition and an intention. You could get a flash on something that turned out to be absolutely true but the moment that happens, you've now influenced that future because you have information about it that's influenceable. So it becomes very tricky to know what you are actually dealing with. That's why I like to bring these things into the lab so you are able to separate out, "Is it this or is it that?" in a controlled fashion.

You talk about our deep interconnectedness; is that another way of saying we are entangled in a field of consciousness? We are not simply interdependent but rather we are connected in different ways, including psychic ways. So it's deeper than say an ecological interdependence, almost like an ecology of consciousness. *Do you think that that kind of connection and awareness consciousness lasts after death?* This goes into my pile of "I don't know." I am aware of the evidence. Many of my colleagues are absolutely convinced that consciousness persists in some way. I am not personally persuaded that the evidence means an individual personality will continue after death. Mediums claim to talk to disembodied personalities and we know from experimental tests that mediums can receive verifiably correct information. But I don't have sufficient imagination to be able to understand how if a consciousness did persist, in order for it to remain a personality, it would have to retain a large chunk of the memory of that original person because memory is the thing that makes a personality. If you don't have memory anymore, you are just an awareness. This suggests that disembodied spirits would have to retain a large chunk of memory and that raises lots of questions: We tend to think of "spirits" as independent entities, like separate people. But practically everything we know about psychic effects suggests that separate objects and absolute independence are illusions. If instead everything is connected and beyond the usual ideas of space and time, then why should there be separate spirits? Is there just one "big" spirit? *One mind can have many parts.*

Or memory from past lives? Yes, and future lives because you are not in space or time anymore. Now we are speaking about beings outside of space-time unless you adopt the idea that there are multiple sheaths of bodies, as in kind of a chakra model. Maybe that's true, maybe that's not true. Some of my medium friends say that if a person dies, they step out of the physical body and into another one, which does have something like memory of the recent body. Therefore ghost-like things can persist for a while and then fade away. Some form of awareness might persist and maybe that is what mediums connect with. *Russell Targ's* daughter communicated a message in Russian about 20 years after her death.* This problem of interpreting what is going on has been a problem since the beginning of parapsychology. No one has cracked this nut to know for sure what's actually going on. Even though some of the evidence is definitely intriguing, I would say the jury is out: We just don't know.

You've studied people connected to sensors who see slides generated by computers of things that are arousing, which clearly seem to show that the future bleeds into the present, suggesting that we can read what's going to happen in the future. So that means that time isn't linear. Somebody once quipped: "Time is God's way of making sure everything doesn't happen at once." I think it's more than an illusion; it's a real thing, but its true nature we don't know. But that is also true for virtually everything in physics: We don't actually know in a fundamental way what a photon is. We don't know from first principles what energy is. We don't know what a magnetic field is.

What we have are some very good mathematical models and methods that describe how these concepts behave in certain circumstances. A particle physicist might argue and say, "A photon is made out of quarks." But when you actually get down into the underlying elements of what a quark is supposed to be, it's an extremely abstract form of mathematics. A mathematical model is a symbolic representation of something. That's all it is. And that's one of the reasons why I wrote *Real Magic,* because esoteric magic is based on a symbolic representation of reality. If you manipulate the symbols, you manipulate reality itself. I don't know how many physicists think about this, but the parallel is there and it seems awfully close to me.

Let's go through your books to see how they evolved. How did you get from The Conscious Universe *to* Entangled Minds? I wrote about 80% of *The Conscious Universe* when I was at Princeton University. One of the complaints I'd heard most often from the faculty was that none of the psychic experiments had ever been replicated. I knew that wasn't the case because I actually read the literature. So I decided to write a book that described how meta-analysis has been applied to psi phenomena. I wrote it as an academic book and no publishers were interested in it, or perhaps they didn't understand it, so I set it aside. Some years later, when I was at the University of Nevada, I was featured in *The New York Times Magazine* in 1996.[2] The next day, I had four offers from publishers who said, "Would you be interested in writing a book?" I ended up accepting a contract with HarperCollins and was able to finish the remaining 20% of the book.

The next question after that book was, "Now that I am convinced that empirically something's going on, how do I explain any of this?" I decided to take the physics route mainly because even sympathetic people said, "This seems to violate what we know about physics." I knew that wasn't true either so that led to my book *Entangled Minds*. Then people who read it said, "So maybe it doesn't violate physics and it's repeatable, but how do we understand it?" If you try to take a materialistic perspective, it gets really difficult to try to understand it, so I decided to look for clues elsewhere. The first route I took was the Eastern esoteric traditions, because I know that in classical yoga with the tradition of the *siddhis* (yogic powers) although many Westerners haven't heard of the siddhis. So that was the origin of my book *Supernormal*.

After writing that I decided to tackle the Western esoteric traditions, which we know a bit better because it's part of Western history, and as such perhaps a little more compatible to the Western mind. This involves esoteric traditions like Gnosticism, Hermeticism, the Kabbalah, the Rosicrucians—all part of the same Western thread. *And Theosophy.* Yes, and Christian Science and Mesmerism. At the earlier stages of these traditions there's a strong overlap with Eastern esotericism. Basically, it's the same until about a thousand years after Plato. In *Real Magic* I make the analogy that technology is to science as magical practice is to the esoteric traditions. That is, magic is an application of the esoteric worldview, just like today's technologies are an application of the scientific worldview.

Supernormal focused on classical yoga and the Indian sage Patanjali, who wrote the *Yoga Sutras.* One-fourth of Patanjali's book was devoted to a description of psychic abilities and ways of developing them. So at least 2,000 years ago, yogis had no problem with the idea that special powers (that we'd call psychic) existed and if you meditated long enough, and do certain practices, you will eventually gain those powers.

Many people have read the Autobiography of a Yogi *by Yogananda where siddhis were observed in more recent times.* Yes, the idea of the siddhis is not that they are only ancient legends, but rather that they are ever-present and available in trained minds. Most of the elementary siddhis are essentially a form of clairvoyance, or perception of the past, the present and the future. That's the very first siddhi that Patanjali writes about, because it arises so often as a result of dedicated meditation practice. We did a survey at IONS where we asked over a thousand meditators what kinds of experiences they had encountered as a result of their practice. About 75% of meditators reported that they had one time or frequently experienced psychic abilities. It took a few years to find a journal that would publish that paper because our findings went far beyond the idea that meditation is only good for relaxing your mind and body. We were asking, what else happens? The answer is that you start perceiving the universe, just as Patanjali wrote about long ago.

They experienced precognition and levitating—what other siddhis did they experience in contemporary times? Roughly 25 siddhis are mentioned in the *Yoga Sutras.* If you categorize them, they mostly fall into classic forms of clairvoyance and precognition. People describe having these experiences roughly in correlation with the way that they are described in the *Yoga Sutras.* If you imagine that the Sutras are described from elementary to very rare, that's more or less what you see today. The number of people who say that they've actually become invisible or levitated, which are part of the classical siddhis, is extremely small. Hardly anyone can obtain those abilities, but something as simple as seeing through space and time is considered so vanilla that practically anyone who meditates long enough will experience those phenomena. They might call it a synchronicity or a precognitive dream, but it's so common as to be no big deal. *Yogis in India have levitated.* In *Real Magic,* I devoted a chapter to write about historical cases of levitation that are quite persuasive, but also rare to the point of perhaps one in a billion people able to do this.

What's your next book about? I know I have at least two more nonfiction books in mind, then I am going to write a science fiction novel. I'm currently collaborating on a screenplay for a television science fiction series.

In the IONS lab are you mostly working with the intentional changes in plasma? I have four ongoing mind-matter interaction experiments involving physical targets in one form or another. The idea is to mentally affect the direction that the plasma is flowing. The goal is to mentally pull it towards you, or to push it upwards, or push it to the right. Plasma is a nice target because it's considered the fourth state of matter. It's not liquid, solid or gas; it's something quite different. It is also exquisitely sensitive to very weak electrical fields. Even the tiniest bit of a stray electric field will affect the movement of the plasma. When you look at the plasma, it almost seems like a living system, it's so dynamic. It's an undulating electrical wave. So far we have pretty good evidence that the plasma's movements conform to what people mentally want it to do. *I tried with a plasma ball but don't have the subtle measurement devices to see if my intention had an effect.* I am looking at a variety of other kinds of physical systems too, including water, with the hope that some of them will be more responsive to mental intention than, say, an RNG.

We know the observer influences the outcome. Yes, if somebody does not want to see an effect, or is unconsciously afraid of the idea that their minds can push the world around a little, then that becomes their intention and they won't get any results in the experiment. So in these experiments you need people who are in alignment with their conscious and unconscious desires and beliefs. After doing these kinds of experiments for many years, I've learned to guess fairly accurately which kinds of people are likely to perform well at these experiments and which will not.

It would be interesting to see if you got better results from people who meditated just before doing the experiment so that they were in that kind of neutrality. Yes, although we generally ask a participant to try one of these experiments for only about 20 minutes. It's difficult for most people, even experienced meditators, to maintain the attentional states that are most effective for mind-matter interaction tasks. It requires a curious combination of "effortless striving." This is an extremely high state of motivation, with a super concentrated focus of intention, but simultaneously with no effort at all.

It's so easy to be pessimistic when we think of what we are doing to the environment and we have autocrats being elected around the world. Is there hope? I am a chronic optimist, so of course there's hope. If you look in historical terms, we've been in much darker places in the history of the United States and the world. We've been on the edge of an apocalypse many times so there's plenty of room for hope. *Scientists say we have around 11 years to turn around global warming and then we are devastated, so how do you have hope in the face of what we are doing to the planet?* Humans are very resilient and the planet is much more resilient than we usually think. It has all kinds of feedback systems that we hope will prevent it from doing something bad. If the environment changes quickly, which it seems to be doing now, evolution will respond to that as it has done many times before. In terms of our current civilization, we might lose half or far more of the human population. But I am reasonably sure that something will persist.

The thrust of evolution is not a linear path, it's more like a spiral, but ultimately it does seem to be moving in a more orderly direction. So I think the Earth will persist very nicely. Whether it continues to be a host for humans, as we know ourselves today, is another issue. I feel sure that there's intelligent life that permeates the universe in many different forms. So from a universal perspective, the direction of the intelligence arrow is almost certainly upwards. While humans are resilient, we can also change very quickly (in evolutionary terms) both in our behavior and physical structure according to what the environment demands. In the short-term, humanity will very likely face some major problems. But in the longer term, I think we'll be okay.

What do you do for fun outside of research in the lab? Working in the lab is the most fun I have. I find it exhilarating to explore the edge of the known.

Books

The Conscious Universe, 1997
Entangled Minds, 2006
Supernormal, 2013
Real Magic, 2018

Endnotes

1 https://www.oxfordmartin.ox.ac.uk/blog/automation-and-the-future-of-work-understanding-the-numbers/
2 https://www.nytimes.com/1996/08/11/magazine/they-laughed-at-galileo-too.html

Roger Nelson, Ph.D.

The Meaning of Global Consciousness

Photo by Manfred Weber, Neuss, Germany

Questions to Ponder

The Global Consciousness Project organized by Dr. Nelson has found significant effects in data that should be random, when millions of people are unified in their emotions responding to great tragedies or in mass celebrations--but not in hugely exciting competitions like World Cup Soccer. What are the implications of these findings?

Dr. Nelson refers to Teilhard de Chardin's idea of the noosphere and physicist David Bohm's idea of a virtual pre-material informational Implicate Order as theories of ultimate reality. This could explain how the anomalous structure arises and how distant healing works. Implications?

I was born in Nebraska on a farm close to a town called Broken Bow in March 1940. *You're Pisces--do you identify with the two fish?* In a jokey sort of way. My colleague Brenda Dunne and I worked together for many years and she was also a Pisces. We said, "We are not wishy-washy. We are flexible."

What kind of influence did you grow up with in that farm family? What was their religious belief and what influenced you to become a psychologist? It was a farm family only for the first five years of my life. We moved first to a tiny village and then to the "metropolis" of North Platte, Nebraska. With a population of about 15,000 it was the fifth largest city in Nebraska. So, I am a bit of a small-town boy. My parents were spiritually organized around morality, but nobody was a big churchgoer except my oldest brother who was crippled from birth and found church a very good social place to be. My religious background isn't conventional, focused by parishes or churches, so much as a kind of curiosity about the spiritual aspects of life. I was eventually able to develop that focus into my own version of spirituality. It's basically the same as my parents: Try to be a decent person and make sure you pay attention to the Golden Rule. I think also about the Golden Braid, which is my way of recognizing that there is a commonality to all spiritual traditions that has to do with spirituality and not religiosity. It's more about our personal relationship to each other and to the world. It has to do more with ethics and morality than with beliefs about afterlives and religious traditions.

How did you get from Nebraska to Vermont and then to New Jersey? I was a good student and got a scholarship to the University of Rochester in New York. I left the small town where I grew up and connected to the larger world by way of Rochester and then New York City. I went to the military and was stationed in Germany, where I became a little more worldly and cosmopolitan, and met my wife-to-be. I learned another language and culture and traveled around Europe to expand my horizons. I came back thinking I would be a photographer, but I wound up as a graduate student in experimental psychology at New York University and later at Columbia.

I took a teaching job in Northern Vermont in a small college and I moved there with my wife and son who was then three years old. After eight years I became discontented--this was in 1980--because it seemed the college was working toward becoming a job-training diploma mill rather than a liberal arts institution, which is my idea of college. By accident, I saw an ad for a job in Princeton, which interested me. I met the people and we liked each other. My combination of skills and attitudes led to me getting the job, which was coordinating research in the PEAR lab (Princeton Engineering Anomalies Research) founded in 1979 by Robert Jahn at Princeton University.

What have been your most difficult challenges in your life and how did you deal with them? I've lived a charmed life and haven't had to deal with tough challenges. That's of course not exactly so, but the truth is I have been quite fortunate. I have been healthy and I've never been arrested for being black, for example. Basically, I am a white, middle-class person with the kind of credentials and background that allows me to get jobs and make a living. I have been fortunate enough to fall into things that suited my skills and my personal mission in life. One chapter in my new book, *Connected,* is titled "Design by Coincidence." I have a long history of things falling into place because it seems they were supposed to be a part of a sequence.

For example, when I started the Global Consciousness Project (GCP) in 1997, I wanted to have a beautiful picture of the earth as a logo because I thought we human beings were working toward covering it with a *noosphere,* a sphere of intelligence made of coherent human thought. I found a wonderful website called Fourmilab, run by John Walker, who is a consummate technologist. He created Autodesk, the corporation which provides the best computer software for CAD drawings and tools. Coincidentally I discovered that John was hosting an experiment that I had helped some other people set up a few years earlier. When they couldn't run it anymore John took it over. It was an online RNG experiment where anybody in the world could try to change the way random numbers would flow. I got in touch with him and he was immediately interested in the developing GCP. He felt it could be something like a super-conducting super-collider for parapsychology and he started building software for the project.

Meanwhile, my son, Greg, who is also a consummate techie and excellent programmer, coincidentally had a period of time when he wasn't overloaded with commitments, so he had time to create the architecture for the GCP in 1997 and early 1998. Although software changes every three months or every year in the world that most of us know, Greg and John's software has been running perfectly for 20 years, from 1998 until now. You can see the results at the GCP website (global-mind.org). *Can people play with RNGs themselves?* Yes, that is possible. The GCP network isn't for that purpose, but if you go to fourmilab.com you can find the online experiment.

A lot of scientists find that if they are working on a problem, they wake up with the solution or they dream about it. Have you found that you access information in that non-local way? I've had a couple of dreams that were relevant to the Global Consciousness Project. In a dream, I received a big box, which came with instructions that said to push a button. I pushed it and a flap opened and a thin transparent tube rolled out while the box started making groaning sounds. I could feel that there was something difficult going on. Pretty soon the clear plastic tube expanded as a furry mass passed through. It eventually popped into the receiving tray and I could see it was the paw of a very small lion, a baby. This was a practice session for birthing a lion.

It was exactly at the time I was setting up the logistics for the GCP working through a number of possibilities for people who would take care of the money. I talked to a group that said, "Oh that's going to cost a quarter of a million dollars and I don't think we can handle it." It turns out we haven't spent that much in 20 years. Basically, the lion was telling me that this project is going to require you to buckle up and work at it. *It sounds like labor pains.* Absolutely, that's exactly what it was.

Also, I use the *I Ching* every once in a while, especially when working together with other people to solve problems. The *I Ching* has never failed to give wonderful advice whenever I've used it, which is maybe half a dozen times over the last 20 years. The guidance at that time was to be sure that I was on a correct path, a righteous path, and to be sure that anybody I contacted would respond in kind.

Many scientists that I've interviewed meditate. Is that something that you include in your life? Yes. My first contact with this whole area at the frontiers of science was in Nebraska when I was a high school kid. I wanted to learn Judo, or some kind of martial art, so I went to the local library. They didn't have any books like that, but I found a book on yoga written by an Indian and a German. This combination of Western and Eastern understanding of yoga was a beautiful introduction for me into that path and the book included some meditation. My deeper introduction came through a retreat in the mid-90s with Thích Nhất Hạnh, a Vietnamese Buddhist monk. For many years, my meditation has involved what I learned from him. Essentially, calm yourself, sit quietly, make sure that all is well and then relax into understanding. "Breathing in, I calm my body, breathing out, I smile. Breathing in, I see myself as a flower, breathing out, I feel fresh." Part of the meditation is simply breathing; that's probably the deeper part, relaxing, being calm, and trying to get out of my own way. *It's a challenge to calm the monkey mind, but are you able to quiet the mind when you are meditating?* Not easily and not often but I don't worry about it and that calms the monkey mind quite a lot.

When you started working with RNGs, how did you know about them? William Tiller was working with them at Stanford, was there any connection between him and you on the RNGs? I started working with random number generators (which can be thought of as high-speed electronic coin-tossers) at Princeton in 1980. We began with commercial sources of random numbers used for cryptology and necessary random source work in industry. Later we started building our own machines. Tiller's work was done later, probably in the early 1990s and it is different. I've never tried to imprint a kind of consciousness state into an electro-mechanical device as he does. I learned about RNGs from our local engineers, and from Helmut Schmidt, a physicist who worked for Boeing aircraft. He wondered whether RNGs might be susceptible to influence from consciousness. He was at first by himself, but eventually other people started picking up on the questions. We were in a sense replicating his work at Princeton.

You are a psychologist working with engineers. How did you get together? We were a very diverse group, the classic interdisciplinary team. We had a couple of physicists or astrophysicists, an electrical engineer, a philosopher and two psychologists in the group when it was at its largest. I am also a kind of on-the-job trained engineer. I was interested in physics and chemistry before I was interested in psychology. I was a physicist first, then I became a psychologist, and because my family history included things like building stuff, I was always kind of an engineer. Brenda Dunne once said, looking at John Bradish and York Dobyns and me, as we were repairing and refining an experiment that somebody else had built, "I see a pit crew." We worked together like a pit crew does. Somebody puts the tire on, somebody fills the gas tank, somebody makes sure that the windshield is clean.

Princeton prides itself on being an Ivy League prestigious scholarly university. Did you get flak from colleagues who questioned working with intention on machines? The remarkable man who started the PEAR lab, Bob Jahn, didn't have much time left over from acquiring funding and defending the lab from the sometimes very active oppositional forces within the university. He told us, "You guys work, I will take care of this." He did, and the lab maintained a functional, active, productive presence but he had to sacrifice the time in the lab getting his hands dirty, although he loved that aspect of research. He had a lot on his plate, as he was also a major figure in academic engineering. His textbook on plasma physics is still the go-to textbook for that field.

What surprised you about the findings? What did you find in terms of what human intention can do to change the sequence of zeros and ones? What has surprised me most about the 25 or 30 years I've been working in this area is that it shows a persistent small but real effect. Nothing we do in the experiments ever seems to make it go away. One of the best practices for people working in difficult research is to do everything you can think of to make the effect disappear. If you can't, it's probably real.

If I sit and read a book in front of the generator, totally focused on my book, do I still have an impact or do I need to be thinking coherence, coherence, coherence? Probably it's much better to just forget about it and read the book. From my experience, everyone is different. Some people may do well if they work at being focused on the task of changing the numbers, but most people do better by setting the intention and then relaxing and getting out of the way. It's a little bit like the meditation, calming the monkey mind that throws a monkey wrench into the process of connecting with other people or with RNGs, for that matter.

I read that two people in love have more influence than two people who aren't entangled sitting in front of the machine. Yes, that's one of the experiments that we did in the PEAR Lab which I think was profoundly informative. The effect size for the two people we call a "bonded pair" is better than either can achieve by themselves. I think what happens is that the bonded pair knows about connection; they have practice. You're trying to connect with this box sitting on the table which is producing random numbers. If you make it somehow a part of you, it is more likely to absorb the information you are trying to provide to it in the form of your intentions to get high numbers or low numbers.

How did this expand to be a global project? We did experiments in the lab for about ten years with individuals trying to change the behavior of these machines and with enough success we thought it was time to help other researchers with replications. So, we designed smaller, portable RNGs. We eventually built them small enough that you could put it in your pocket and take it out into the field to examine group consciousness. We went to concerts, to churches, we went to rituals--all kinds of interesting gatherings. That work led to other questions--about what effect distance would have and whether we might profit by using multiple RNGs. Eventually those questions matured into the GCP, designed to monitor events in the world that bring large numbers, often millions, to share thoughts and feelings, using a network of many RNGs.

I was amused that you went to academic gatherings and they didn't show coherence. Right, an academic conference isn't something that produces a lot of coherence, nor is a street corner or a train station likely to produce any patterns in the data. But if you take an RNG to a religious celebration or most any traditional ritual, the natural resonance often produces patterns in our data. We were looking at a "consciousness field" produced by a group of people who are resonant with each other and the context. They generate coherent, structured information that gets absorbed into the RNG bit-stream and becomes the pattern that we see in the data.

The GCP idea really began to jell in another one of those wonderful coincidences. I was at Esalen Institute in Big Sur, California, in the hot tubs and met a young couple who were on a mission to organize a global meditation called GaiaMind.[1] I was there with colleagues thinking about distant mental interactions and was in a deep meditation on group consciousness and speculating about Teilhard de Chardin's idea of the noosphere, the coalescence of human thought.[2] I heard the GaiaMind organizers talking about setting up an experiment exactly like I envisioned for a test of the global consciousness idea. So, I asked all my friends with RNGs, "Please take data during this meditation and send it to me." We got a surprisingly big difference from what's expected during the time of the global GaiaMind meditation.

Not long after that, Princess Diana was killed and that became another prototype for the GCP. I asked my friends to take data during her funeral. Because estimates indicated two billion people would be paying attention, it was a very clear and obvious gathering of global consciousness, focused and coherent because of the circumstances. The results again were strong, with 100 to 1 odds that you would get such results by chance. That provided encouragement to make a permanent listening network to ask the question whether global consciousness might really exist in a nascent form. I gathered resources and my son Greg got to work, collaborating with my friend John Walker. I talked with Dean Radin,* Dick Bierman, and a few others about experimental designs. By August 1998 we were able to begin taking data with a functioning network of RNGs asking what happens if a huge event brings people to a global focus. A few days after the network was set up we had our first test when two American embassies were bombed in Africa, shocking the world.

The network provided continuous data collection, 24/7 and month after month. Our general hypothesis allowed monitoring of many kinds of events like tragedies, terror attacks, accidents, natural disasters, and also more positive events like the Olympics and World Cup Soccer, and celebrations like New Year's and the Kumbh Mela in India. We gradually homed in on a rigorous protocol and procedures for selecting the events that we would look at. We knew we would have to repeat the same experiment again and again in order to pull the signal out of the noise.

You have the formal data from 500 events. Are the RNGs still collecting data? Yes, I am still looking at events occasionally when there's something that appears to be important, such as a horrifying natural disaster like a tsunami that kills thousands. *What were the other big effects? The World Trade Tower attack was also really big.* Yes, it was huge. Early on the 11th, the data started changing and they continued to be divergent for three days from what's expected of random numbers. At the time, there was nothing else like it in three years of data.

The fact that it started having an effect before the planes hit, does that mean that our sense of time is not linear in that the future bleeds into the present? Or was it the emotionality of the attackers? There are two major possibilities and I think you hit them both. One of them is that there were about 50 or 100 terrorists and the leadership who knew about it. The 19 people who died flying their planes into the towers were saying their final prayers. Normally we wouldn't see much happening in the data, no matter how focused and concentrated 50 people are; rather, we more likely see a correspondence when millions of people share emotions. So, my preferred explanation is that just as human beings individually have precognitions or premonitions about the future, a global consciousness may also have a premonitory capacity. We've looked at that same question in other data; in 2006 one of my colleagues, Peter Bancel, gathered all the earthquakes that had a Richter magnitude of 6 or greater. He found about 100 on land, and one of his exploratory analyses showed a V shape of the RNG data which starts about eight hours before anything big shows and bottoms out just at the time of the main quake. The 500 big quakes that occurred in the oceans showed no patterns in the data.

Which caused the biggest effects? It's a difficult question in part because we are dealing with statistical outcomes and those are always estimates with big variability. The nice thing about repeated experiments such as the GCP formal series of tests is that even though each individual experiment may only have a small deviation, when you add them up, it becomes impressive. Though the average effect size is modest, the 500 events in the global consciousness formal data sequence add up to a composite z-score of 7.3, more than 7 standard deviations from expectation. That's one in a trillion odds. If you did this experiment a thousand billion times, you might get a result like that by chance. There isn't any doubt that there is a real effect here. Now it's time to start figuring out how and why and what it means.

What are some of the other big effects? The tsunami in Japan, the Haiti earthquake, any tragedy that kills lots of people that grabs our attention. At the positive end of the spectrum, the Kumbh Mela may have 30 million people trying to bathe in the Ganges to wash away their sins. Those are the ones we've looked at and they tend to add up to a very impressive degree. New Year's celebrations are a regular event where we all become focused and share feelings like love and appreciation. These are powerful emotions that bring people together and make a difference. When Nelson Mandela died, there was a powerful effect, as there was when Michael Jackson and Steve Jobs died. In all these cases there was a huge outpouring of emotion on the part of at least hundreds of thousands, probably millions.

What about the 2016 presidential elections with the surprise victory for Trump? It was flat, not distinguishable from the wandering random trace I talked about; it didn't ring the chimes of the GCP. In another example where effects are surprisingly small, a lot of people mention World Cup Soccer matches, but it's more complicated than just making a lot of people pay attention. There are always two sides: There's Brazil and there's Argentina, so we have a competition rather than a collaboration or cooperation. We have looked at a lot of sporting events, especially World Cup Soccer because it is huge in terms of the number of people paying attention globally, but they are paying attention in a competitive spirit rather than this spirit of love and compassion that seems to be so powerful in the GCP data. That may be what happened in the election of 2016. The election in 2008 was a very different story, which produced a powerful deviation in the direction we predict.

Tell us how the noosphere fits in? The GCP has demonstrated a tiny effect with outsize implications. People have differing reactions. Some don't believe the effects we see are real, but those who know what it's about mostly agree that our science is good, and the results indicate something important about human beings. The idea is that we have some work to do moving toward our destiny as a kind of protective sheath of intelligence for the earth, what Teilhard de Chardin called the noosphere. We are in the process (many of us believe) of figuring out conscious evolution--how can we rethink our lives and help each other get on with the business of being a unified intelligent species. The data and analyses we produced are scientific support for the poets and wise ones who have been saying we are all one. What we see in the data indicates that we are all connected at a deep, unconscious level. My belief is that if there is a growing understanding that it is the case, we will be able to make this idea of interconnection more conscious and make it a powerful part of our lives. That will help us change some harmful behaviors like exploitation and mindless consumption so we can do a better job of being the evolutionary top-dog.

What does this tell us about quantum field theory and how human emotion can affect the coherence of a machine? There are quite a number of interesting and potentially useful models or theories, but none of them are very easy to test so we really don't know. The most appealing to me is derived from David Bohm's work where he describes the Implicate Order as a kind of background out of which everything arises. He talks about a pilot wave as active information that allows the virtual particle or the virtual connection to become real. The process is an actualization of the virtual to manifest in the concrete world. We are basically a structured collection of bits and pieces of reality, all of which were once virtual.

The important thing is that we are a source of information or consciousness which is information in the sense of structure. If we think about something like healing, a healer is a source of structured information for a person who needs the healing. In Bohm's model he says that the actualization occurs when it's needed. If there is a need for this particle to come to exist, then it can be actualized from the implicate background as a particle in reality. And in a similar way, a wounded or ill person presents a need for structure that the healer has in his or her consciousness, even from a distance. Active information may be the medium that connects them. Maybe such connection is why we see that prayer is alive and well, even in our modern times. Humans have been praying from the beginning of time and I don't think we keep on doing things that are not useful. It's a time-tested activity.

Machines don't have any needs, so how is it that a machine responds to human intention? It's an important question and the answers are more in the realm of philosophy or art than science. When we design an experiment, we are setting up an implicit request that the RNG be our instrument for detecting consciousness fields, for example, or for detecting the effects of intention. In a profound sense, an RNG partakes of or exists in the consciousness field arising in the implicate order––some people talk about consciousness as fundamental. The RNG is a remarkably well-designed instrument that is poised in the now, where the future is completely undetermined. The random bits won't exist until a quantum process actualizes as an electron with the indeterminate choice of leaping the quantum barrier or not. This device is capable of absorbing information, we think, in some metaphoric sense at least, so that the sequence of random numbers will be changed to become what we want or what we intend.

Quantum mechanics tells us things exist as potential and whether it's a particle or wave depends on whether it's observed, so everything exists in potentiality unless there's a focused consciousness? It certainly seems so in some cases where you can set up an experiment in a laboratory. The outcome may change depending on whether there is an observation or not. For example, Dean Radin's experiments with the double-slit experiment have people observe or not observe (mentally) the electron or photon as it's passing through the double-slit apparatus. He has done a remarkable job of showing that the consciousness of the observer does change the behavior in this quantum system. *Speaking of intention, Princeton graduation has less rainfall than would ordinarily be predicted because people don't want it to rain during graduation.* That's a study I did where I compared rainfall on the days when everybody is gathered outside for various kinds of events and found there was a statistically significant tendency for less rain, less often than on control days. It is one of my favorite papers; its title is "Wishing for Good Weather."

Do you think the Internet enables us to have a more integrated global mind, so that that could be a tool to think peace or equality? Yes, the Internet and mobile phones allow us to connect with people everywhere, even on the other side of the world. We need to always understand that the material and the spiritual aren't really opposed; rather, they are complementary aspects of the world. Maybe the noosphere, which is what these RNG experiments indicate, is made of nothing but information. The blogosphere and the Internet, and other things that depend on devices and electrons flowing and wireless radio waves, are essential aspects of an emerging global community. If we are able to work toward developing the full potential we have as human beings, the global Internet and the global consciousness may become a high-quality complementary structure enabling us to be one people on the planet.

Does that mean that you are hopeful in the face of global warming and climate change and how radically we are destroying the planet? I am a congenital optimist, but I am frustrated and sad. We've destroyed over a million species in the last few years; it's horrifying. And yet there are signs that a lot of people are waking up. Once you have a well-formed question, its answer is right around the corner. I am optimistic because there are so many beautiful, thoughtful people in groups working toward a brighter future. My friend Stephan Schwartz,* one of the most productive people I know, has spent much of the last several years working on promoting wellness as a way of being on the earth. One of Stephan's books is called *The 8 Laws of Change*. It's about how to be the best human being you can be. Instead of measuring our success in terms of how much money we have, we must learn to measure in terms of how much wellness is created. The difficulty that we must face is how to talk to people on the other side of the fence. We've forgotten how to be in families; we've forgotten how to be good neighbors; we've forgotten how to be compassionate.

How do you create balance and have fun? What's not fun? Like most people fortunate enough to love their work, it is deeply rewarding. But it's not all work. I have hobbies: I do some photography and a little sculpture and build things and fix things and I love that. I love being in a salon, a group of people who are in conversation. You become a bigger, better person as a result of conversation with other people who are, like you, looking for an exchange of ideas that sometimes become profound discoveries.

I have a small men's group and my wife and I have friends who enjoy the same kinds of things. I am very interested in gardening and sustainable living. We belong to a community in Upstate New York called White Hawk, an intentional community that's growing and is now about one-third of its intended size of 30 families. We are all interested in making sure that the land is cared for rather than exploited. Some religions believe that we are intended to have dominion over the earth, which I think has been terribly misunderstood. What it should mean is responsible husbandry and caretaking. This is the attitude needed if we're to survive and prosper.

My 2019 book documents the GCP project and provides access to the results in a way that I mean to be transparent. The GCP website has the information but it is too complicated--the basics are buried in the details. It is deep and thorough, as is needed for documentation. The book is entitled *Connected: The Emergence of Global Consciousness*. It tells the story of how and why the project started, how we put it together, and what we have learned scientifically. At the same time, I wanted to talk about why this is important and what we can do to manifest our destiny as keepers of the earth. I include art and poetry to help understand what noosphere really means and to urge everyone to envision and help create a vibrant future.

I have another recent book, published early in 2018. It is co-written with my friend Georg Kindel, who publishes a beautiful magazine called *OOOM* in Vienna. The book is in German, with the title *Der Welt-Geist: Wie wir alle miteinander verbunden sind*. That means *The World Spirit: How We are All Connected*. It covers the GCP but also a wider territory of new scientific work showing the remarkable capacities of human minds.

I'm interested in how we learn to mature as humans. How do I live up to my potential as a human being? How can I make it possible for other people to step onto the same path? We had a sign in the lab that said, "If we all work together, we can totally subvert the system." Even if you are at the top of the heap in economic terms, why wouldn't it be preferable for you at the top of your heap to look around and see beauty instead of sadness and tragedy and chaos? We all must come to this soon. It is essential to the creation of beauty and life in an interconnected world.

Books

(Ed.) *Research in Parapsychology,* with D. Weiner, 1987

Der Welt-Geist: Wie wir alle miteinander verbunden sind, Vienna, Austria, 2018

Connected: The Emergence of Global Consciousness, 2019

Websites

http://global-mind.org

http://noosphere.princeton.edu

Endnotes

1 http://www.gaiamind.com

2 https://teilhard.com/2013/08/13/the-noosphere-part-i-teilhard-de-chardins-vision/

Gary Schwartz, Ph.D.

Spiritual Psychology: Talking to Other Dimensions

Photo by the Picture People

Questions to Ponder

Dr. Schwartz was raised to be an agnostic. What events led him to research communication with departed spirits and angels? What technologies is he using?

What does super synchronicity indicate to him?

How does viewing consciousness as primary change how scientists understand their research?

What is self-science?

I was born in New York, June 14th, 1944. I was raised in what I call an ultra-Reform Jewish home. My parents were so Reform that they virtually never went to temple and I didn't participate in any synagogue activities. However, for the grandparents, I was bar mitzvahed at age 13 and I was taught how to pronounce the words in Hebrew. However, I still don't know what the words mean. As a result, between an upbringing that was really areligious, and growing up in a science-minded Western culture, I was essentially an atheist. I assumed that the Bible was mostly a myth and that spiritual ideas were superstitions based on misperception, such as the sun revolved around the earth. It was quite a slow and semi-painful process for me to discover that the materialist perspective was actually incorrect, as established by the totality of scientific evidence.

I was playing music professionally by the time I was 13. I started with clarinet and then graduated very quickly to rock and roll (Motown) and then jazz (funk) guitar. I even played at my own bar mitzvah. Music still is very much part of my life as I consider myself to be both an engineer-type and a music-type person. Both of those qualities contribute to my academic work and my sensitivity over the years.

It takes a lot of courage to do this kind of new science. What was it about the family that you grew up in that encouraged you to be an independent thinker and to get into Harvard? I've done many interviews over the years but no one has ever asked me the question in exactly that way. Was there something in my family, in my upbringing that not only encouraged me to be creative and open-minded but also encouraged me to be brave? The creativity part is simpler to explain because my mother was a serious musician. She studied classical piano at Juilliard but never graduated because she got married and pregnant with me. Later she went back to school and became a grade school teacher.

My father was also a very creative man. Besides playing the piano, he was a scientist and an engineer who loved chemistry. He had a joyful love for chemistry. They both loved music and they were both very loving, passionate people who encouraged me to be creative. I was given a lot of freedom, probably too much freedom where I spent hours in the woods. I loved animals and I had multiple swimming pools with snapping turtles, water snakes, painted turtles, carp and other fish, rats and many other kinds of creatures. My parents tolerated a lot of critters. I had chemistry, physics, and electronics labs in the basement and my parents encouraged me to play that way along with my guitar.

The interesting question is about courage. My mother had very strong opinions; if she had a passion, she went for it. If her child was not being given a fair opportunity, she fought for him. Something my mother modeled was standing up for what she believed in and pursued it vigorously with passion. My father was a Ph.D. student at Columbia University, who left engineering, completed a second undergraduate degree in pharmacy, and ultimately owned a little pharmacy. He worked seven days a week, often 10 hours a day to build his business. Until he could afford a part-time second pharmacist, he had to run the store. And although it was not his passion, nonetheless he did it, with responsibility and grace. My parents modeled strength and I can honestly say that I was blessed to be encouraged both for creativity and for courage.

How many siblings do you have? I have one younger brother for whom academics was not his cup of tea. He went to a few years of college and then got into his father-in-law's business running a charter fishing boat. He is happily married, has three wonderful children and multiple grandchildren. In some respects we are as different as day and night. *Siblings have a way of sorting out you are going to be this and I'll be that.*

Why did you decide to study psychology? I was in first year electrical engineering at Cornell University and within a few weeks discovered that I didn't quite fit in, so I transferred into arts and sciences and became premed. My original plan was to be a philosophy major and a chemistry minor, and then go to medical school to become either an anesthesiologist or a psychiatrist––two different medical specialties that care about reducing pain and suffering. However, I happened to take an introductory psychology class and an abnormal psychology class and then discovered the field called psychophysiology, the relationship between mind and body. This was a relatively new field at the time, represented at Cornell by a young assistant professor, Harold Johnson. He applied for his first NIMH grant in 1964, which involved placing electrodes on people's bodies, measuring brain waves, heart rate, blood pressure and so on. For example, he presented emotional slides and then measured people's psychophysiological reactions to them. I discovered that that particular field integrated philosophy, electronics, chemistry, biology, physiology, psychology and medicine. I was especially interested in psychosomatic medicine, placebo effects and so on.

I ended up becoming a psychology major, chemistry minor, premed, with an emphasis in psychophysiology. I decided to do my Ph.D. in clinical psychology with a psychophysiology focus and then go on to medical school. Due to a strange sort of circumstances, I ended up at Harvard because I originally went to the University of Wisconsin. Upon completing my Ph.D. at Harvard, I was so successful that I was offered Assistant Professorships at both Stanford and Harvard. Consequently I decided to postpone going to medical school to continue my research. I was soon offered a tenured professorship at Yale. My plan was to go to medical school there but I realized that the research was flourishing so I decided that it was easier to collaborate with physicians than to become one.

You had an experience with your first wife where you heard a voice that protected you before an accident on the highway when it said, "Put on your seat belt." Was that part of your coming to understand Spirit with a capital S? No, I considered it "anomaly" and "paranormal," so I gave it zero thought. I was deeply grateful for the experience that saved both of our lives, but it didn't in any way consciously affect my thinking.

It's only in hindsight, having worked with multiple evidential mediums, where I've been forced by totality of the data to conclude that not only are they getting accurate information from formerly physical people, postmaterialist beings, but they are also getting information from higher sources, higher beings—including the divine, universal consciousness, God, Allah, whatever word you wish to use. I now look back at that event and say, "Wow, I actually experienced something that was a very precious gift and I wish I had appreciated it sooner." In the accident, which totaled our car, I lost complete memory, save for those few moments prior to the accident. Psychic John Edward is very clear when he said, "Thoughts pop into my head and I learned how to interpret them. I don't experience them as my thoughts. I am not generally in those thoughts but they are using my thought apparatus." *That's how I experienced it too. A client's father recently died and came to me, saying he was irritated because he wanted to move on and their grieving held him back.*

What led you to start studying psi phenomena? I didn't study the paranormal at all. I was a mainstream psychophysiologist studying mind-body-medicine and health psychology. I later expanded into energy healing approaches from a biophysical perspective, as opposed to a paranormal perspective. My journey began because of two things: One was a theoretical discovery that occurred when I was a professor in Yale which forced me to become open to paranormal phenomena, including survival of consciousness after physical death. I wrote about this discovery in *The Living Energy Universe.* It's a mathematical theory about how memory is stored in all systems that have feedback loops and how that dynamical memory can continue as an info-energy system after the physical system is no longer present. Just like the light from distant stars can continue long after the star has "died," the energy information from a dynamical system can manifest feedback and can continue in "the vacuum of space" as well. That mathematical analysis led me to become open to all kinds of paranormal phenomena, but I didn't go near it experimentally for over ten years.

Then I started meeting people who had lost loved ones and deeply wanted to know whether their loved ones continued after they physically died. The first person was a former research and personal partner, Dr. Linda Russek, who grieved the death of her father, a very distinguished clinical and academic cardiologist. I also met Susy Smith who wrote 30 books in the field of parapsychology and life after death. She became my adopted grandmother and used to call me her illegitimate grandson. Susy was fond of saying that she couldn't wait to die so she could prove that she was still here.

Stardust Johnson was the wife of a beloved professor of music at the University of Arizona who was murdered by a drug dealer. Stardust really wanted to know whether life after death existed scientifically because she said it would give her a reason to hope and a reason to live. So, I was pushed there by scientific theory and evidence as explained in *The Living Energy Universe* (discussed below) and I was pulled there by Linda's love for her father, Susy's love for her mother, and Stardust's love for her husband. When you are pushed and pulled in the same direction, it's very hard to resist.

Originally, I started doing life after death research in secret; I metaphorically experimented with my left pinkie (i.e., on the side) to explore it. As the research progressed, it became my left hand. I began to meet evidential mediums and do controlled experiments. It became my right hand when I had so much evidence that I could no longer deny the evidence.

In one of your books, you talked about God as kind of an intelligence, order and helpfulness. Does that relate to consciousness? The book is titled *The G.O.D Experiments,* where G.O.D stood for Guiding, Organizing, Designing process. This is similar to what Larry Dossey calls the One Mind. This is also a Buddhist concept—the idea that there is a universal consciousness, the "ground of being." It's infinite and it's beyond space and time in terms of its capabilities. In religious circles this consciousness is called God, Yahweh, Allah, and for Native Americans, the Great Spirit. Other terms are the Creative Intelligence or Source, but they all reflect this infinite universal conscious and creative intelligence.

How is consciousness different than a quantum information field evidenced in non-locality? Consciousness refers to awareness. It refers to experiencing and has intention. That's more of a mechanics of an expression of the organization of the consciousness than it is "the consciousness" itself. When you look at the primacy of consciousness concept, it says that those things are themselves expressions of this infinite potential rather than the potential itself. This is simple logic. If you ask me am I a scientist first or a logician first, I would say I'm a logician first. Reasoning takes you to methods, which you use to make discoveries. Our capacity to put things together, to combine reason and intuition is part of what we mean by consciousness and extends to the concept of the soul.

How do you know that it has the quality of emotion like love? First of all, we get there by reasoning. So, what do we mean by love? We can describe love as an experience when we are having it; we know it. We can also describe love as an intention. Usually it's an intention for kindness, caring, protection, nurturing, empathy, compassion, empowerment--it's a very rich and complex set of emotions. We can describe love in terms of its behaviors and how we literally physically treat people. We don't know that you and I have the same experience of love but we know it propels us and motivates us to do similar things (e.g., express kindness). Once we understand that love is an inference in everything else other than ourselves, then we can ask to what extent do other things in the universe show love.

One of the first people to discover this concept was Sir Isaac Newton who developed the theory of gravity, which was very novel in his day. You have a mutual attractive force which is holding everything together, which is completely dependable and trustworthy as it brings everything together. Does this sound like two people who have a mutual attraction and are brought towards each other and can depend on it? We infer that this mutual attractive process is an expression of love. Love brings people together in a mutual way. Ideally, it's trustworthy, it's constant and it grows. What Newton realized was that his idea of gravity as a mutually attractive force was dependable, trustworthy, unconditional. It didn't matter whether it was a big planet or a little planet, it was a completely unconditional mutual attractive force. Newton reasoned that this universal attractive force was an expression of God's love in the universe. I conclude that of all the creatures that we know, humans can be thought of as infinite love machines.

If we look around the world, it looks pretty dismal. Why are we so dastardly when we are created from this higher consciousness? My reasoning process is we are a microcosm of the macrocosm, as explained by Deepak Chopra and Menas Kafatos* in their book *You Are the Universe*. The universe gives us opportunities to make choices. It's an opportunity-providing God, which means that we have the gift of choice accompanied by responsibility for our free will. The truth is if we abuse the planet, we and everything else are going to pay a price. I was raised in an ultra-Reform Jewish household. One of the things that Jewish people have asked is how could God have allowed the Holocaust? My sense is God says, "You want freedom, you have to take responsibility. With one comes the other." We have horrific things happening in the world right now, no question about it, as well as the greatest opportunity to really grow up and make changes.

People who have had NDEs have these transformative moments and scientists who have had mystical experiences are now seeing a new way. Postmaterialist science expands science where ethics and spirituality can be reunited with science. If we can create a practical SoulPhone, you will be able call your deceased grandfather, or you will be able gain wisdom from Albert Einstein or Jesus himself––presuming they choose to take your call. If we could have a real collaboration between here and There, people wouldn't fear death and would see all this as an opportunity for hope, which might provide the inspiration that we need to come together to make the kind of hard choices we have to make as a maturing species. What I find most compelling, the evidence that pushes me as a scientist to infer the existence of a loving and caring ultra-intelligent, ultra-creative universal consciousness, is what I call super synchronicity.

You wrote Super Synchronicity: Where Science and Spirit Meet *about that. Do you think it's fair to say that synchronicity where two uncoordinated events come together indicates a higher guidance?* If it was just a pair of events, which I call type 1 synchronicity, I would say no because it could be just a coincidence. But when you have six or more events, which I call the type 3 synchronicity or super synchronicity, also called serial coincidence, it becomes "too coincidental to be accidental," as Susy Smith used to say. *Is that like your experience with panthers?* I've seen a whole series of many things. One that comes to my mind, because it's so acutely silly, are ducks. It's so ridiculously improbable and yet when you look at the combination of it, the only other explanation is some sort of intelligence is having great fun and giving feedback to someone who's willing to listen to it. When you have six or more of these events, it becomes so improbable that you realize there is a level of scripting, a level of creative coordination that makes you become very humble and very grateful. A total of 19 synchronicities occurred with ducks. [*The duck story is found online.*[1]] *Do you think it's a trickster kind of playful game?* I wouldn't call it trickster, I would call it a very clever sense of humor and this intelligent consciousness will be as humorous as the person with whom it's attempting to communicate. I've known hundreds of people who have had this kind of super synchronicity, so I have a lot of experience in this area. When I started exploring this further, I was given a profound lesson.

I took part in a mediumship development circle led by a woman who was a very gifted medium who channeled a group she called The Council of 12. I asked her to ask the Council about the ducks and tell me if this has any meaning. They said, "If it walks like a duck and it talks like a duck." All of a sudden I got it. The way *The Afterlife Experiments* book ends is with the statement that if the totality of the evidence for life after death, metaphorically, walks like a duck and talks like a duck, at some point you are willing to say it's a duck (i.e., evidence for life after death). I realized the same thing applied to super synchronicity and inferring consciousness. When you really try to analyze the sophistication of what it would take to have all this come together at this particular moment in a receptive individual, if it walks and talks like a super intelligent mind, then it probably is.

What else does this revolution in science involve besides a recognition that consciousness is primary? It changes the way you do experiments as it leads you to recognize that your expectations can be interactive with the experiment. *Quantum physicists have told us that the observer and measurement changes whether it's a particle or a wave.* Yes, quantum physicists inferred that consciousness was part of nature but they didn't make it primary. The average physicist rarely studies the impact of the mind as playing a potential role in what's occurring in their experiments. There's a whole range of scientific areas where consciousness could be playing a part in the results. The second area is that what we call parapsychology is no longer *para;* it's mainstream (as James Carpenter* explains). It's actually reflecting more the essence of the universe than an anomaly that is outside the realm of science.

In terms of mind over matter like placebo? What we call placebo is no longer placebo; it's literally the effects of conscious intention whenever you are reflecting and interacting with the physical world. Einstein's General Relativity was a bigger theory than Newtonian gravitational physics. Therefore Newtonian physics became a special case of a bigger theory. You could use Newtonian equations quite well but it's not the correct explanation, according to Einstein. You need a vision where time and space are not fixed. If you start approaching the speed of light, Newtonian physics literally doesn't work. But under many conditions, it works in practice.

Post-material science is not anti-material; it's just bigger than material. In many areas, science can go on doing what it's doing because it works but that doesn't mean that it's the correct explanation. It will never make the required predictions and will never allow for other phenomena unless we expand our theoretical realms. So the new science embraces all of science, but it's reframing it to have a bigger perspective. This means that we can make sense of data that doesn't fit and we can make new predictions that can be tested, confirmed or disconfirmed. It is an expansion of science that reverses and changes the way you experience what you know.

It adds another dimension which is the observer effect? In the process of adding that dimension, it transforms the way we think about the lower dimensions. Einstein's theory is the bigger theory but Newtonian math works within a certain range of relativity. So it's not just an additional dimension; post-materialist science goes further than that. It says that we are going to have to reframe the way we think and teach, for example, about what the brain is capable of doing. The brain becomes much more of a brilliant antenna-receiver-transmitter system than it is a creator of consciousness. We'll make more discoveries about the brain by seeing it more correctly––that's the vision. *I read about a hospital in Arizona where a woman was having surgery; her brain was dead physically but afterward she described the operation.* There are many thousands of cases of people having experiences during surgery where they validate certain aspects of what was taking place.

Your latest work is with the SoulPhone. Where are you in the process of working with these devices to communicate with spirits on the other side? We are in stage 2 of prototype development. Stage 1 was to get a replicable effect to demonstrate proof of concept, a working device, but not a practical one. The metaphor is a Wright Brothers moment. When the Wright Brothers demonstrated their plane, the first flight lasted all of 12 seconds and the longest flight was 59 seconds. It was by no means practical but it showed that flight was possible. Prototype level 2 was to get reliability to the point where we could do live demonstrations, so we could show that an individual hypothesized to be collaborating with a post-material person in spirit (from now on "spirit helper") could use this technology and use questions to verify that it really is a specific individual.

Using a switch to indicate a yes or no? Yes, using electronic technology. The purpose here in working with a research medium is to get confirmation with the technology that our spirit helpers are actually who they say they are. The third step requires substantially more funding to go from a working demonstration that's proof of concept; let's say 24 minutes to answer 24 questions. A spirit helper could give a yes/no answer and we would know whether it was accurate or not in two or three seconds as opposed to many minutes.

What technology are you using? We are looking at multiple technologies; one that we're having the most fun with right now, partly because it's very cost effective, is to take advantage of a phenomena that's well known for over a hundred years. It became commercially available for the general public in the 1980s as plasma globes. It has a Tesla coil in the middle developed by Nikola Tesla. When you generate voltage, it produces streams of light. If you touch the glass, the light will be attracted to your finger. (You can buy them from the Smithsonian Museum or Spencer Gifts.) They are exquisitely sensitive but you can only see this effect when you do sophisticated computer averaging.

(Dean Radin is also using this tool at IONS. I tried with my plasma ball but don't have the measurement devices.)* The question becomes, what if we have one of our spirit team put hands in or on the globe? Using very sophisticated averaging techniques of tens of thousands of images, we can uncover effects in the shifting of the light that you can't see with your naked eye. We've been doing this research for over a year.

Why can't you use a regular switch and have the spirit helpers signal yes is up and no is down? It takes energy to affect the physical world, but when a person "dies" they lose most of their physical energy. You need a super sensitive sensor where you can simultaneously filter out all the unwanted noise––so you need hypersensitivity and hyper-shielding. Today we have technologies that can detect single photons of light and measure electron flow to one to the minus 15 zeros. We have the capability to detect their energy and so they can use it to communicate. But if you can record such low levels of energy, you must also have super-shielding, super-filtering, and that's expensive. The sensitive side is becoming less and less expensive, but the shielding part is really a challenge. Engineers have not only developed robotic capability to make these tiny devices, but they have created "clean rooms" where the air is made ultra-pure, the temperature is controlled and the environment is electrically shielded. We don't need to wait for new technology; what we need to do is to find people with vision to use that technology and bring the resources together to integrate it for this purpose. I concluded that the probability of creating a practical SoulSwitch is the same probability that we had for developing airplanes and cell phones.

You mentioned on one of your websites that the project has been helped by Albert Einstein, David Bohm, Thomas Edison and Nikola Tesla, as well as Susy Smith and Carl Sagan. What do they want to convey to us? When my wife Rhonda's mother died, which happened before she met me, Rhonda discovered that she was able to get evidential communication from her mother. She subsequently published a book about her journey called *Love Eternal* and developed her ability to be a selective evidential research medium. We have brought all of this into the laboratory. Susy and Rhonda are very loving, very smart women who really care, not just about their families, but about humanity. Our project is ultimately not about the technology, it's about the transformation of humanity as the veil becomes more permeable. Just as we take for granted that we can fly around the world and now fly into space, people will eventually take for granted that the veil need not separate us from the greater reality.

What's an example of something that you've learned from one of the spirit helpers that you didn't know? I started out as a skeptic, raised by my parents to believe that all of this was impossible. Though I am philosophically an agnostic (i.e., I try to approach anything new from the vantage of "I don't know"), I was educated to believe that none of this was possible. I had to give up those educated disbeliefs and open my mind accordingly. I read *The Unobstructed Universe* by Stewart Edward White (1988). He claims that just as light can pass through a window and electromagnetic waves--radio waves--can travel through walls, so too can formerly physical persons who function like electromagnetic waves.

I am first and foremost a scientist. There are certain claims that I have heard from mediums, for example, that I suspect are probably true--for example, that spirits can be in more than two places at the same time. However, until I conduct controlled experiments, I must say, "I don't know." I've not done formal experiments to test this hypothesis (yet) but it's easily testable. Apparently for spirits, they can "bi-locate" or "tri-locate." *In* Autobiography of a Yogi, *Paramahansa Yogananda describes living yogis doing that.* For the purpose of spirit communications, I can place a plasma globe inside a light-tight enclosure with a webcam to measure the momentary changes of the plasma streams. We can invite a member of our spirit team to take his or her hand, put it in through the box and glass to create physical-like plasma change effects. The fact that I have some difficulty fully accepting this fact, that's my problem. Part of this whole post-materialist revolution in science is recognizing that nature and physics are far more fluid than we currently understand. It wasn't until I actually performed the research, tested the technology, and found that it worked, that I was forced to conclude this is real. Until I can validate something with empirical science, I have to say, "I don't know." I am really excited about conducting future research using two completely automated plasma globe systems located in two places (e.g. Arizona and Ohio). We could invite Harry Houdini, for example, who has demonstrated replicated effects with individual plasma globes, to be in both places at the same time and use both pieces of equipment at the same time.

What other research institutions do paranormal research? You head the University of Arizona (UA) Laboratory for Advances in Consciousness and Health. You are fortunate that UA is sympathetic partly because you bring in the money, but a lot of universities probably wouldn't allow it. Yale would probably not have permitted it. It's still somewhat of an embarrassment at UA, but because the UA is in the southwest, in a somewhat more spiritual environment, there is more collaboration and more integration here, it's a friendlier environment to do this kind of work. I think of the Canyon Ranch Resort as being the Harvard of Health Spas and one of the most sophisticated integrative medicine-type health resorts in the world. I had the privilege of being the Corporate Director of Development of Energy Healing at Canyon Ranch for 11 years. They were very supportive of both our afterlife science research and our energy healing research. The Windbridge Institute is a private institute that was created by a former postdoctoral fellow of mine, Dr. Julie Beischel,* to study mediums. As far as I know, my laboratory is the only university-based laboratory in the world that is doing research that focuses on life after death and developing technology to detect the presence of spirit. At the Institute of Noetic Sciences, Dr. Dean Radin* is the head of their research division. They mostly do more "conventional parapsychology," which I do not do. Mediumship and SoulPhone technology are better described as spiritual psychology than as parapsychology.

You helped start a new institute for post-material researchers. We created the Academy for the Advancement of Post-Materialist Sciences. It's modeled after the National Academy of Sciences and the Association for the Advancement of Sciences. It's a small institute of like-minded individuals focused mostly on education. Its purpose is to provide a place where individuals who are interested in the transformation of science from materialism to post-materialism can find a home (www.aapsglobal.com). We have edited two books in this area. I am the senior editor of a 2019 book called *Is Consciousness Primary?* It's a collection of essays by founding members of AAPS describing how they came to the conclusion that consciousness is primary. Many (but not all) parapsychologists are "materialists" who see consciousness as created by the physical world, by the brain, for example, and therefore they infer that when the brain dies, consciousness must die. Their vision is an extension of conventional materialism as opposed to a post-materialist perspective that sees consciousness as fundamental if not primary in the universe.

The primacy of consciousness hypothesis sees consciousness as preceding information and energy which in turn precedes matter. That means that the umbrella concept for all of science is consciousness: not energy, not information, not matter, but consciousness. This perspective turns everything on its head. It's like realizing that the sun no longer revolves around the earth. It completely flips everything on its head, and when that happens, science expands accordingly.

You wrote a book about healing modalities. Did you find that Reiki is more effective than Johrei, etc. or does it depend on the person who's transmitting the healing energy? There are no large-scale clinical studies with any specific symptom or disorder that would allow one to determine whether or not one technique is better or worse. Based on substantial data––mine and others'—it appears that some people are better healers than others. We don't know what all the variables are: techniques, motivations, personalities, or their receptivity to learning. We conducted an experiment to see what happened when people had a one-week intensive training course with Rosalyn Bruyere, the energy healer. The students were physician fellows in the Integrative Medicine Program at UA and we looked to see if we could objectively measure their ability to detect energy. On the average, the group was more accurate in sensing energy after the training than before. But some people showed large increases, others showed small ones and some people showed no increase at all.

Have you learned how to transmit Reiki and Johrei and other healing modalities as part of your research on healing? In the early days, meditation, biofeedback, and even energy healing were very controversial, but today they are pretty much taken for granted. Contemporary super-controversial topics include survival of consciousness after physical death and studying higher-order spiritual phenomena like synchronicities and the reality of the Source. In the area of energy healing, I knew that the only way that I could study this was to learn techniques myself. Consequently I learned at least six different healing techniques and I became quite proficient employing a few of them. I learned enough that I could produce real effects—meaning that it comes through you as opposed to being you. However, I actively avoided learning to become a medium. I chose to remain an outsider, metaphorically, like a blind person trying to study color vision. I thought it was really important that I not be seen as being biased in favor of spiritual phenomena so I actively avoided learning mediumship and other intuitive techniques. However, this doesn't mean I haven't occasionally experienced the phenomena.

I have had intuitive moments where I got into the flow and

obtained highly evidential strings of information about people's deceased loved ones. From a purely statistical point of view, the accuracy of the information is so far beyond chance that the only way I could explain it was to say that whatever they were doing, I was doing too. However, I'm not a medium in that I can't see or hear spirit. Dr. Robert Stek, a dear colleague, refers to me as the "Helen Keller of afterlife science." *How do you know the medium isn't reading the sitter rather than the departed spirit?* I have witnessed hundreds of research mediums and I would say that 90% to 95% of the time they don't get the most obvious information. That's completely antithetical to the telepathy hypothesis.

You said that many of the mediums directly work with a guide, so are they channeling what the guide says? I don't think so. *I assume Susy Smith probably is eager to help you with your research? Did she ever say my illegitimate grandson, make sure you do so and so?* Susy knows me well enough to know that I am one of the most difficult people to get through to. She knows the best way to get through to me is to go to Rhonda or Suzanne Wilson or go to Marc Anthony. It turned out that we could predict who would show increased energy detection accuracy using a personality scale called the Absorption Scale. It measures the extent to which people become immersed in a subjective experience. People who are high absorbers are people who are typically more sensitive; they tend to experience sunsets more deeply or they get lost when they read a book. It would make sense that people who are more intuitive and more connected to experience might benefit more from an experiential technique.

You mentioned that women were the flame that propelled a lot of your work in this field, and it seems like most of the mediums that you've mentioned are women. Do you think that's because we are socialized to be more receptive and to be better listeners? Some of it is socialization but I think some of it is nurturing. Women are biologically predisposed to be more caring, sensitive, and empathic for evolutionary biological reasons. That doesn't mean that some males can't be very empathic and it doesn't mean that all women are empathic. *My experience in doing this kind of work is that you have to step aside from yourself and listen. You have to shut off your logical mind and your ego. Maybe women are more able to do that because they are not as socialized to develop their logical minds at the expense of their emotional and intuitive minds.* Yes, I think socialization is an important part of it but there may also be a natural proclivity as well.

You also talked about being a self-scientist. Is that a Gary Schwartz concept? I didn't learn it explicitly from anyone. We can apply the scientific method to the laboratories of our personal lives and learn how to become more "evidence-based" in our own personal experiences. Virtually all of the great mystics were self-scientists; they learned what worked and what didn't work to learn how to better chart their capacities. Detectives like Sherlock Holmes are self-scientists who are constantly learning how to grow as a result of experience and labeling it as science. Learning from experience is one of the ways that we can produce an evidence-based spirituality so instead of science and religion being separated, they can come together in a mutually supportive nurturing way. *Feminist methodology is more comfortable with experiential and folk traditions and diaries and journals so that this ties into valuing personal experience.*

Are angels real? I wrote a book called *The Sacred Promise* where the last quarter of that book is devoted to the Angel Hypothesis. I was brought up to believe that angels were socially accepted myths like Santa Claus and the Easter Bunny. But evidence-based mediums tell me about their communications with angels and I have to give them the benefit of the doubt because they can be highly accurate about the spirit world. The systemic memory hypothesis indicates the plausibility of higher levels of organization and consciousness, so that angels are theoretically feasible. I started meeting people telling me about my alleged angels. I confess this in *The Sacred Promise: How Science Is Discovering Spirit's Collaboration with Us in Our Daily Lives.* I met a person who was doing a postdoctoral fellowship with me in integrative medicine and in energy healing. He claimed to see angels as part of his first childhood memory. He was very science-based, worked at IBM, and was very "sane."

The claim is that virtually everybody has angels. Moreover, if you could learn to connect with your angels, you can work with them to live a more full and successful life. But could the Angel Hypothesis be brought into the laboratory? A medium claimed that my angel was showing up at the lab and deserved to be taken seriously. Also, I had developed the openness to ask the universe for information and allow the information to pop into my head. I decided to initiate that process which I explained in the book where my ego was completely gone and I was hardly aware of my body. I call it a state of "non-sensory awareness" in a state of "pure loving attention and intention."

I once had an experience of a gigantic radiant being that either had wings or energy that looked like wings and it took me completely by surprise. I hadn't had any drugs and hadn't been drinking and I was shocked. When I asked in my head, "What is your name?" and I heard in my head the name Sophia, I thought, "That's ridiculous. There are no angels named Sophia." As soon as I said that, the image disappeared. However, I discovered angel Sophia who was foundational in angelic philosophy historically. In some traditions, Angel Sophia is considered the first manifestation of the divine. Sophia purportedly gave birth to all the other angels and ultimately the material world. In other traditions, the angel Metatron is considered to be the male appearance of the Divine, and the angel Sophia the female appearance (i.e., they are two expressions of the same universal energy). There's a fourth century gnostic text called *Pistis Sophia* about the Sophia tradition, so I realized that maybe I had tapped into something fundamental. [*See the Hurtak* chapter.*]

I wondered what would happen if Susy could bring Sophia to a medium under lab conditions. I described one of the paradigms that I learned through Susy in *The Sacred Promise* book. I call it the double-deceased paradigm––when one deceased person brings another deceased person to a medium under blind conditions. Would the medium recognize an angel as opposed to a physical person? I actually did that kind of research and surprisingly obtained positive findings. The medium didn't believe in angels but she was really gifted in connecting with Susy and getting evidential information. She said, "This person is really big. She is not really family. She is very close to you and she's got blonde hair. I feel like I can't look at her in the eyes. She has the same name as my daughter," which is Sophia.

I also conducted technology research, published in the journal *Explore* in 2011, that addressed the angel hypothesis. We used a low-light CCD $40,000 camera, that was cooled to minus 70 degrees centigrade, originally purchased to measure biophoton energy. [*See John Kruth's* chapter.*] I confessed the secret angel aspect of this research in *The Sacred Promise*. The experiment was completely computer automated. PowerPoint was used to give audio and visual instructions in the middle of the night when none of us were physically present. Half the trials invited Susy to go into chamber and half the trials invited Sophia to go into chamber. Much to my amazement, the effects for Sophia were substantially greater than the effects for Susy and they appeared to be affecting the organization of higher frequency energy which is also consistent with the Angel Hypothesis. Although it's political suicide in an academic world to do such research, I put together a formal proposal with a budget of close to two million dollars. The proposal was called HARP which stands for Hypothesized Angel Research Program. When I directed my NIH-funded biofield research center which included multiple projects, the budget was $1.8 million for three years. If any single discovery could transform humanity and wake them up to a greater spirituality, it would be to give angels a voice. Thus far, there have been no private donors interested in supporting such controversial research.

Some of the scientists I interviewed for this book talk about aliens and reptilian forces that are lower-astral-level creepy beings. What do you think about the negative spirit side? Let's talk about negative people first before we get to negative non-human beings on this planet. What's the probability that the people who were negative on this dimension when they lived in the physical continue to live in the post-physical? I am forced to entertain a hypothesis that survival of consciousness is ubiquitous and it's the nature of the fabric of the universe itself. Consequently, the question becomes less whether negative people can continue to exist in the post-physical, but more as to, how can we be protected from these individuals? I have done virtually no research on this question, but Susy Smith said when she was physically alive, "Don't let people play on Ouija Board," because those risks are real.

What about non-human entities that are more negative? I had the privilege of knowing Harvard Professor John Mack who became controversial because he ended up seeing patients in his practice of psychiatry who claimed to have been abducted by aliens. John came to the conclusion he couldn't attribute all of their experiences to mental illness and developed a deep interest in life after death shortly before he died. He had a secret meeting with me and we discussed his emerging interest in some depth. John told me about a documentary that featured his work with abductees. Another physician named Lynne Kitei wrote *The Phoenix Lights: A Skeptic's Discovery that We Are Not Alone* that I helped get published. She also produced a documentary about *The Phoenix Lights* and I invited the two of them to the UA medical school to screen their documentaries.

What surprised me is that Wikipedia, which tries to be objective, has unfairly attacked a lot of these visionary scientists such as Dean

Radin, you, and Eben Alexander.* I can understand being skeptical but I don't understand attack.* It's even worse than attacks; it's abuse, it's deceptive and lacks integrity. *It looks like whenever there is a new paradigm shift, and you've said this is bigger than Copernicus' discovery, that people resist change. Humans don't like to have their worldview shaken up.* There's an evolutionary benefit to preserving the status quo as a protective conservative stance. The instinct to preserve what you love and what you think is real has certain advantages. *It seems like academics are especially resistant to change although they should be at the forefront.* Yes, science is a very conservative process; it's designed to err on the side of false negatives (thinking that something is not true when it really is) rather than false positives (thinking that something is true when it really is not). Science prefers not to believe that something might be true until there is a sufficient amount of evidence.

Let's look at the themes in your books. What was the thrust of The Living Energy Universe, *1999?* It confessed the theoretical discovery that occurred when I was a professor at Yale, a discovery that I kept secret for 10 years. The notion is that all systems at all levels, from subatomic particles to super clusters of galaxies, have recurring feedback loops. In this process, information and energy are stored and therefore a system, regardless of its level, size, or complexity, has memory. It follows that everything that exists, to various degrees, has memory, "learns" and therefore is "alive." It was mathematics and physics that led me to that discovery. This hypothesized systemic universal feedback memory process led me into all kinds of controversial areas, including the possibility of NDE or OBE and survival of consciousness after physical death.

The Afterlife Experiments, 2002, was the first book where I discussed research with mediums including both single-blinded and double-blind experiments. I was forced by the totality of the evidence to conclude that not all claimant mediums were frauds and that the best explanation for the observed accuracy in genuine mediums was survival of consciousness. The 2005 book *The Truth About Medium* was inspired in part by the appearance of a popular TV show at that time; *Medium* was created by Kelsey Grammer after meeting a woman who he met through me named Allison DuBois. He decided to do a fictional TV program very loosely based on her life, as described in her book *Don't Kiss Them Goodbye.*

The G.O.D Experiments book was published in 2006. *G.O.D* stands for Guiding, Organizing, Designing Process. The subtitle is *How Science Is Discovering God in Everything, Including Us.* This book is based on work that I did trying to understand the nature and origin of order. I wanted to know if the idea of randomness could actually explain evolution and the emergence of everything we know. I was compelled to the conclusion that even "randomness" itself can't occur by chance. The presence of apparent randomness requires a pair of conditions not present in nature in order for "true randomness" to occur. I was inspired to write that book after working with the "Dream Detective" Christopher Robinson in England. He developed the capacity to see the future in his dreams and then use it in police work. He worked part-time as a secret agent for British intelligence and Scotland Yard. In the process of working with him, I was forced by the totality of the evidence to conclude that patterns of information were occurring in the research (and therefore in our collaborative lives at the time), which could not be explained as random. I was compelled to conclude that Christopher was receiving sophisticated communications in his dreams that could not be explained unless one entertained the idea of a universal mind or One Mind.

The Energy Healing Experiments, 2007, was based on a summary of research that my colleagues and I performed as part of an NIH Center for Frontier Medicine in Biofield Sciences, funded by the National Center for Complementary and Alternative Medicine. It addresses the question, is energy healing real? How can we bring energy healing into the laboratory? Can we study its effects in cells, on animals, on people? The book is filled with compelling evidence and applications. *The Sacred Promise* (2011) was focused on bridging self-science and laboratory science to demonstrate how the two of them can complement each other. The book examines some of the more visionary and controversial aspects of life after death. It raises questions about whether we can build collaborative relationships with beings, including higher beings, and whether we can verify such relationships scientifically.

I had the privilege to write a book with Paul Davids called *An Atheist in Heaven*, a book about Forrest J Ackerman, who was a luminary in the sci-fi industry. In fact, "Forry" as his friends called him coined the term sci-fi. Forry was a devout atheist who was convinced that life after death and God were not real. However, Forry was also a very caring person who said, "I am going to come back if life continues." After Forry died, very strange things began to happen to Paul, such as physical alterations of documents, and he made a documentary about it called *The Life After Death Project*.

Do you think that in the face of such a short window of time we have to keep the planet from imploding from global warming that we can develop this kind of partnership and turn things around? Yes, I am convinced that that is a very serious possibility, partly because more and more people who are on the other side are caring and asking, is there some higher purpose? In the extreme case we could annihilate large segments of the world, including ourselves through nuclear war or excessive climate change. If you take a broader perspective that there is no death, there is only transformation, then the dinosaurs are extinct in physical form on the earth, but in the greater living energy universe, they continue as living energy beings. As long as photons exist, as long as information and energy exist, then they will continue to exist in their essence and they'll continue to be "evolving." The research strongly indicates that there's no death, there's only transformation. One ends up becoming more hopeful for the species even though we engage in tremendous self-destruction and cause each other tremendous pain and suffering. In terms of the short-term, yes, I am very concerned. But from a longer term and larger picture, I am not only hopeful, I am even encouraged.

In the face of your tremendous productivity, how do you find time for fun? It looks like you collect sculpture and Native American art? I've been collecting Native American art since I was a junior in college. Rhonda happens to share this passion and our house is filled with mostly Pacific Northwest Coast and Southwest Native American art, as well as African Art. When you don't have children, art becomes your children, so to speak. We help support artists and, when it's time for us to transition, the art will be donated to others who can appreciate it. *What else do you do for fun? A lot of people said nature is really important to them.* Our backyard is like a nature preserve, we live right up against the mountains and we have no fences. We have many bird feeders, lots of deer, and herds of Javelina. I used to read mysteries and was an avid reader. Rhonda and I both love music and so we enjoy seeing young people develop their skills. We also enjoy artistic and playful cars.

Some people say it takes a couple of marriages to finally arrive at one where you are mature enough to have it work. Do you feel that's true? I would have to say the answer is certainly yes in my case. I've been privileged to have been married three times. Each person was special. In the first two marriages, we didn't quite match, we weren't able to give each other what we needed and we weren't mature enough to have made the best choices. I deeply respect the people I was privileged to be with before Rhonda, and I expected to never remarry again. I decided that someone as complex as myself required someone who was equally complex; we would require a unique set of shared interests and values, as well as a shared respect for each other's "space." I would not have been mature enough for someone like Rhonda at an earlier age and I suspect the same thing for her. Some people say that three is a charm and for us we've been blessed that way.

What's your next book? I am completing a book titled *Extraordinary Claims Require Extraordinary Evidence: The Science and Ethics of Truth Seeking and Truth Abuse.* This book complements and extends my new book with Alan Bourey, a lawyer, titled *The Case for Truth: Why and How to Seek Truth.* Humanity will never thrive, or even survive, without a deep commitment to truth seeking, where truth is defined as accuracy of information and understanding. Spiritual evolution grows hand in hand with accuracy evolution.

Is there anything you'd like to add? I've lived my life feeling gratitude and blessings because most scientists have not had the opportunity to grow in a spiritual way at the same time they were growing as research scientists. I've been fortunate to have many experiences with "real magic" as Dean Radin names it. A good 90% of the sculptures and art in our home each have meaningful stories. Most of them have descriptive names and they are reminders of synchronicity and special events that happened in our lives. I always type out the word "Blessings" to end my e-mail messages; I always type it so I remember to express and feel it. With some friends, I sign EGG, which stands for Eternally Grateful Gary. The notion of being able to experience gratitude in bridging science and spirituality is a very precious gift.

Books

The Living Energy Universe, 1999
The Afterlife Experiments, 2002
The Truth About Medium, 2005
The G.O.D. Experiments, 2006
The Energy Healing Experiments, 2007
The Sacred Promise, 2011

Super Synchronicity, 2016
An Atheist in Heaven (with Paul Davids, 2017)
Greater Reality Living (with Mark Pitstick, 2019)
Is Consciousness Primary? (with Marjorie Woollacott, 2019)
The Case for Truth (with Alan Bourey, 2020)
Extraordinary Claims Require Extraordinary Evidence, 2020)

Endnote

1 https://visionaryscientists.home.blog/2019/08/22/gary-schwartz-ph-d-19-syncronicity-experiences-with-ducks/

Stephan A. Schwartz

Explorations in the Non-Local

Photo by Kathy Kenney

Questions to Ponder

What led Stephan Schwartz to spend five years reading the Edgar Cayce readings of people's past lives, etc.? What mechanism does he thinks allowed Cayce to obtain that information?

What experiences suggested to him that life continues after death?

As an experimenter, how did Stephan Schwartz avoid the decline effect? What were the outcomes of his experiments with PK and remote viewing? How does Associated Remote Viewing work?

What trait is most likely to produce a skilled remote reviewer? What have they predicted about the future?

You were born in Virginia, you live in Washington, you've lived in Los Angeles, Washington DC, NYC, etc. Do you find regional differences that impact how people view consciousness? I was born in Ohio and when I was 16 my family moved to Virginia. If you mean do different regions have different cultures, I would say, yes. Paradoxically the more religious the community the less open they are to non-local consciousness as a scientific reality. It's a paradox since all religions at their core are about non-local consciousness, which they call spirit but, when you say in scientific terms that spirit exists as a scientific reality, they balk. This all traces back to the Council of Trent in the 16th century, when the Roman church took anything having to do with consciousness (read spirit) as their domain, and left space-time to science with the admonition that if science got into consciousness and the church could get hold of the scientists who crossed the line, they would kill them. From that came the taboo of materialism and the depiction of non-local consciousness as scary and demonic.

What was your family's attitude towards religion and spirituality when you were growing up? I didn't appreciate this until I was in my 30s, but one of the great gifts my family gave me, for which I am profoundly appreciative, is that they didn't care about religion at all. They weren't atheists but they had no interest in the dogmas and formalities of institutional religion. They did care about consciousness in a very pragmatic way. Mine was an entirely medical family; my mother was a nurse, my father a physician, and uncles and aunts were physicians, nurses, and hospital administrators. My parents understood that consciousness played a big role in the outcome of illnesses or surgeries, but this interest had nothing to do with religion. We never went to church, never talked about it. Thus, I was not saddled with the limitations on thinking that religious dogmas impose; I never had to climb out of that hole.

You said they were aware of consciousness; do you mean as in the placebo effect they were aware of the mind's influence on body? Yes, that's part of it. They were aware that the consciousness of patients, even the consciousness of the doctors and nurses treating the patients, had an effect on the outcome of an illness or a surgery. They witnessed it. To give you an example, as an anesthesiologist my father, after a decade of assessing surgical outcomes in the 1940s, began a practice uncommon at the time, now quite ordinary. He always saw his patients the night before the surgery to explain to them exactly what they were going to go through and that they could have confidence he would see them through. It made a huge difference to people. At that time it was considered bizarre for an anesthesiologist to go around and see patients, but my father felt very strongly that it was important that his patients saw him as a person. His confidence would help their confidence, and his humanity would help their sense of care. Now it's commonplace but at the time it was considered very strange.

So one of the things that you inherited from your family is the ability to have the courage to do what other people think is very strange research? Yes, and also my family had a deep appreciation of the processes of science. So, what I learned as a child was a deep appreciation for the rigor and processes of science and a respect for the idea of consciousness as a prominent factor, but without any particular ideology or theology.

Where did you go to school? A private boys' school and then to the University of Virginia. I was in the Echols Scholars program, an honors curriculum modeled on the Oxford Cambridge teaching form with two tutors and 12 boys. We mostly read the classics and studied philosophical thought. I wanted to be a writer so that worked well for me. Then I went to work for *National Geographic*. I was already writing and publishing—I sold my first magazine piece when I was 15—so the job was a good fit. I got drafted as a medical corpsman in the Army. I had considered becoming a doctor at one point so it seemed a good fit, and I didn't have to shoot anyone. I liked what I did but, over time, I decided that wasn't the way I wanted my life to go. When I came out of the service my father was terminally ill so instead of going back to Washington I went to help my mother at my family's farm in rural Tidewater, Virginia.

Years earlier I had helped my father start a registered Angus cattle breeding operation about which my mother knew very little and dealing with that on top of her grief was too much, so I stayed with her. I got a job at *Daily Press* in the morning and the *Times Herald* in the evening as an investigative reporter and feature writer. I did that for about a year and a half while I was also getting us out of the cattle business. Through a serendipitous connection I was asked to co-write a movie for Colonial Williamsburg, which turned out successfully. That changed my life. A producer at Screen Gems read our script and invited me to come to New York to work as a screenwriter.

After I had been in New York about eight months, I got to be good friends with a well-known art dealer, Sam Green, and through him I met Truman Capote who invited me to a party on Fire Island. It was the beginning of a turning point. Outwardly my life seemed successful. I was 23 years old; I worked in movies and had just purchased a wonderful 190 Mercedes convertible. But inwardly I had begun to think there was something asleep within me. I was missing something important, but I didn't know what it was. As I walked back down the hall, I looked into an antique Italian gilt mirror for a long moment and said very spontaneously, to my own surprise, "You are becoming an unattractive person. Your values are screwed up." I left the party, sat up all night on the beach thinking about this. In the morning I went back to Manhattan, packed up my 190 and drove back to Virginia.

By then I was in a great funk; it is the only time in my life I've ever really been depressed. I just didn't know what to do. I had gone to New York, I was a screenwriter, I had done all the things you're supposed to do, I had a beautiful girlfriend and I had gotten to know all these famous people. Yet I knew I was doing something wrong; I just couldn't figure out what it was.

I had been at the farm for several weeks, when one afternoon I was sitting on the porch that looked out over the bay. I looked up and there was a couple walking down a crepe myrtle alley. The property was at the end of a seven mile school bus road and our lane was nearly a mile long, so it was not a place you got to casually. I certainly didn't expect to see middle-aged strangers walking in the gardens, particularly strangers dressed as if they were on Fifth Avenue in Manhattan.

When they saw me looking at them, they came over, but instead of introducing herself, the woman said, "Do you believe in reincarnation?" I responded, "I really don't know very much about it but I think I probably do. It seems very symmetrical." I asked, "How did you get here? Why are you here?" She replied, "I had a dream that told me to come up and invite you to the Edgar Cayce Foundation. Do you know who Edgar Cayce is?" When I said I didn't, she explained he was a psychic who went into a trance and could get information on anything. The whole conversation struck me as deeply weird but we introduced each other; they were Ed and Paula Fitzgerald.

She told me, "In the dream I could see where to turn to get here and we drove it. I came here to ask you, would you like to meet Thomas Jefferson?" I had gone to the University of Virginia, founded by Jefferson. I could not think how she could know this as there were no computers, no Internet, and asked, "Is he back?" She explained that Cayce said reincarnation was real, that he gave a reading for a boy who said he had been Thomas Jefferson. She went on to tell me something of Cayce's history. What I didn't realize in that moment was that my life had just done a 180 and I was on a new course.

About a week later I visited ARE (Cayce's Association for Research and Enlightenment). As we walked into the library, I saw shelves along one of the walls filled with green loose-leaf notebooks and was told they were the transcriptions of what Cayce called readings. Today I would call them remote viewing

(RV) or non-local perception sessions. At random I pulled a notebook off the shelf. It was a reading given in 1936 for a woman who Cayce told that in a previous incarnation she had been a member of the Essene community. From his description I recognized the location as Khirbet Qumran, and he told her she had been a teacher of astrology.

You know when they say your hair stands on end? The last thing that I had done for the *National Geographic* was research for an article on the Dead Sea Scrolls. I knew that in 1936 nobody knew they existed or that Khirbet Qumran was an Essene community, before the scrolls were discovered in 1947. A young Bedouin shepherd boy following his flock walked by a cave and chucked a rock in. He heard it go clunk and went into the cave to see why and found the urns containing the scrolls. As I read the notebook in the ARE library, I knew that everything Cayce said was correct, but unknown at the time he said it. How could Cayce possibly have known things 11 years before the scrolls were found and archaeology discovered there was an Essene community there. In that moment, I woke up and changed the course of my life at age 24. I called back to the farm and told them, "This is going to be take more than a weekend." It actually took almost five years.

I moved into what had once been a hospital; the ARE in those days rented out a few rooms. I met Hugh Lynn Cayce, the elder son and Gladys Davis Turner, Cayce's lifelong secretary and archivist, and decided that I was going to read all of the readings, about 14,000 of them, as well as the extensive documentation as to their accuracy that Gladys had assembled. I would start at the first one they had on record and read them all in chronological order, and I did. I saw patterns that developed over the decades. Gladys had consulted with archival experts as to how she should do it and became a very sophisticated archivist. As a result the Cayce archive is the largest body of non-local perception, RVing, if you will, that exists and they are meticulously documented.

Reading them transformed me because it gave me a very different idea about the role of consciousness in reality and ultimately it gave me a number of insights that I used to design the protocols for my own experimental work. Seeing how Gladys had created the ARE archive taught me about the importance of very meticulous archival work.

All of this stood me in very good stead when I started doing my own research. I also began reading other people like Helena Blavatsky, George Gurdjieff, and Rudolf Steiner. The ARE had a big library and there were also several universities nearby whose libraries I could use. After I had been at it for two or three years, I decided I needed to read the formal literature on consciousness research, so I started studying the parapsychological journals from the beginning of the 20th century up to that time in 1966.

Did Cayce read the Akashic records, a universal hard drive of people's experiences? Akashic records, Temple of Records and such terminology is for some people a culturally comprehensible way of describing what in science we would call non-local consciousness. Cayce was able to become non-locally aware as successful RVers must do. He was unusually gifted, but because there was no one like him he developed a very idiosyncratic way of accessing the non-local. He went into trance, I think, as a form of disassociation, just as was true in the Delphi Oracle or the Talking Idol of Ix Chel. He was born in the post-Civil War era to a white racist father in a largely fundamentalist Christian community, in a state devastated by the Civil War. He had an eighth grade education. One of the things that people find difficult about the Cayce readings is that they are couched in quasi-King James English. I think he spoke that way because the Bible was the one book that he read over and over. For him it was the language of spirit.

I learned that for the last 14 years of his life he could give the readings without going into trance, but he continued to do it that way because that is what people expected. He went into a sleep-state altered state of consciousness, I think, because he said things which in his community would have been considered heretical if not outright Satanic. He went to sleep so it wasn't the personality of Edgar Cayce speaking; it was his "higher self." He gave the non-local informational domain a culturally appropriate name, a "temple of records." They open to the underlying information structure, like doing a Google search. If you look at the near-death research, you will see very few Christians have NDEs where they encounter Buddha. We have experiences that present information to us in culturally appropriate ways. Eben Alexander,* for instance, had experiences which were consistent with who and what he was.

Does the idea of a collective unconscious help explain how the readings are done? In modern terms, it begins with Adolf Bastian in the 19th century and his book, *The Psychic Unity of Mankind,* in which he described what he called the elementary thoughts of humanity. That had a huge effect on the early German physicists who were the fathers of modern physics: Max Planck, Albert Einstein, Wolfgang Pauli, Edwin Schrödinger and Werner Heisenberg. Carl Jung was also greatly influenced by Bastian, who gave him the idea for the collective unconscious. Also, it's where Joseph Campbell got the idea for the mono myth of the hero's journey. The idea is that informational architectures exist in the non-local domain that are created by multiple acts of intentioned focused awareness. These acts of awareness create non-local memes that shape culture [*see Susan Blackmore's* chapter about memes*]. We're dealing with an information phenomenon but the big mystery question for me is what is information? I don't know the answer and nobody I am aware of knows the answer either.

Some people talk about an information field. Yes, but a field requires spatiality. If you're outside space-time, there is no spatiality. The most interesting part of physics that I'm following is information theory, because they are struggling with trying to figure out what information is. We're looking at information that is manipulated by intention-focused awareness (IFA). Max Planck, the father of quantum mechanics, is correct when he says consciousness is causal and fundamental. Space-time arises from consciousness, not consciousness from space-time. So that's your second definition. Consciousness is causal, physical reality is manifestation. The third is all consciousness is part of the network of life and there is a linkage, possibly quantum, between the individual and the collective. We're part of an information network, a sort of non-local Internet. We live in an informational matrix manipulated by consciousness.

The research on this is overwhelming; there are at least 10 or 12 protocols that each have odds greater than one in a billion of happening by chance. People don't pay attention to it because it doesn't fit their attitude. *What's an example of one of those protocols?* RV, Ganzfeld, therapeutic intention, healing, presentiment. *Can you explain how that is done?* Here's one example. Studies show if you're staring at a blank monitor screen, just a few seconds before an image appears on the screen, your iris will dilate.

The amazing thing to me is that RVers were as accurate about a target that hadn't been selected until after they defined it. Yes, we discovered that as a non-local phenomenon, time meant something very different. It was an information datapoint, not a limiter. RVing the past is looking at an information architecture collapsed to form by IFA and the space-time acts that flowed from that intention. Looking at the present is looking at the existing information architecture of that moment. Looking at the future is looking at the highest probability information architecture yet to be formed by intentioned-focused consciousness at the moment the non-local information is gathered by non-local perception. When one is describing the future one is speaking of a probability. The typical RVing in a lab today would be, "I'm going to show you a picture that won't be selected until after you give me the information. It doesn't exist as a target yet, it'll be selected by a computer out of a big pool of targets." You're describing a probability but because the time frame is an hour, the commitment to show it to you is a compact and the computer is already programmed for time; all of those factors virtually assure that it's 100%.

Are you able to RV yourself? Yes, of course. *Do you use it in your life to view the outcome of this or that possibility?* Sometimes but more often I take things into meditation and open up to options that are available to me. I shape the intention that I'm able to identify, the one that is the most compassionate and life-affirming. If you really want to change the world, go back to the eight laws explained in my book with that title. The key is what I call the quotidian choice. Every day I have lots of choices; of the options that are available to me I commit to always pick the one that is the most compassionate, life-affirming and fostering of well-being. If you tell 10 people you're doing this, you invite them to join you and to tell 10 of their friends, we could change the course of history. Consciousness is causal and fundamental so that manifesting personalities in space-time is kind of like creating an avatar in a computer game, in that the personality is not the fundamental but rather a manifestation of the fundamental.

We incarnate where we can learn and work through our karma? Exactly. *Karma,* a Sanskrit Hindu and Buddhist word meaning action, effect, fate describing what we in consciousness research would describe as non-local information patterns which occur unconsciously through intentioned acts, until you become conscious of them and take control.

Do you think you had previous lives? Were you a scientist? Yes, doing this kind of work. I've had specific experiences of being previously involved with a quest for consciousness but not a lot of details; mostly flashes, little scenes like memories that occasionally will come up. I have a long-time interest in the Cathars because they were a Christian group that believed in reincarnation and gender equality. My sister lives in France, where much of their history is found. When I went to visit her, knowing my interest she took me to a remote Cathar church. I walked into this empty church and, after looking down the nave, I looked down at my feet and I was wearing armor. I was about 26 years old with long blonde hair and a bloody wound in my hand. It lasted only a timeless moment. I understood that I had gone into this church centuries ago as another man. I've had several of those kind of experiences.

What about past lives with your wives or your daughters? I had an experience with my late wife of 30 years, Hayden, in which we saw each other as Vikings. But again, it was just a flash. One of the great gifts I've gotten in my life is I've had two life partners. About two years after my Hayden died, she came to me in a dream and said, "You do much better when you are a partnered person, so I am going to send you somebody." A week later a friend invited me to dinner and said, "Do you remember Ronlyn?" I answered, "Sure, she babysat my daughter Katherine." She told me, "She's come back from England after almost 30 years and she is going to join us for dinner." We started out as friends then began courting and married 10 years ago. I think Ronlyn and Hayden were sisters in the Middle Ages.

When did you begin to do experiments? In 1968 I started with what I called distant viewing. I didn't know anybody else who did that kind of research. I wasn't interested in proving non-local consciousness existed. If I designed the experiments correctly that would just happen. What I wanted to know was how non-local consciousness worked, what could you do with it, and what could it tell us about reality and our place in it? As I read the parapsychological literature, I realized that at that time they didn't really know very much about consciousness and the protocols of that day showed that. The protocols used in psychical research, as it was called then, were not very well designed because one of the principal effects reported was the decline effect. That is, the longer they ran an experiment the worst people got. *Because they got bored like with predicting the Rhine Zener cards?* Yes, exactly. Although J.B. Rhine was a pioneer, based on his experimental protocols, he didn't really understand very much about consciousness. The tasks he asked people to do were incredibly boring. Guessing cards or throws of dice or Zener cards is boring over time and non-local consciousness does not like boring.

I began to think about what kind of experiments could you do that wouldn't produce a decline effect. I had come to think of the Cayce readings as distant viewing because many of the people for whom he gave readings were all across the country or in another country. I knew that he used his senses; he could tell you about smell, colors, and sounds. The key was the ability to attain and sustain IFA and sense impressions so that if I could design an experiment that wasn't boring and turned on sense impressions, I would have a tool that would let me study my questions.

Did the RVers have a psychological profile? In the early 1980s at Mobius [*founder and Research Director of the Mobius laboratory in Los Angeles*] we spent $500,000 and two years testing people assisted by some of the leading researchers in the field to see whether there was a particular profile that was good at doing non-local tasks. We tested over 23,000 people. The main takeaway was that extroverts and introverts approach it from a different perspective. But that's not the defining importance for being able to do non-local task performance. The key to doing it is the ability to attain and sustain IFA. I looked at men versus women, educational level, race. I used the Myers-Briggs and the Torrance Creativity Scale. I found people that score higher on creativity scales do better than people who score on highly rational scales. But the main definer that really stood out was that meditators do better than non-meditators because they develop the ability to attain and sustain IFA.

Is it a learned skill? In part. There are two components: First, your innate ability, just as some people are particularly gifted athletically or at music. Second, your ability to attain and sustain IFA––that's why meditators do better. *It seems like mediums are mostly women but the famous RVers with Star Gate were mostly men because of the military feed-in.* That is probably a cultural effect, although I am not aware if the idea has ever been rigorously tested. The SRI RVers were mostly men because they mostly came out of the service and the researchers were all men in the 1970s.

In my opinion the best group of RVers ever studied were the Mobius viewers. I worked with some of them for two decades, compiling data on them over the years. Gender was not a definer. Hella Hammid, Judith Orloff, Noreen Renier, Fran Farley, Rosalyn Bruyere, and several others were women. But let me emphasize that Ingo Swann, Alan Vaughan, Michael Crichton, George McMullen, Ben Moses, Jack Hauck, and several others were men. They ranged in age from their twenties to their seventies. The ability to focus is the key, which is why martial arts dōjōs, Buddhist temples and Christian monasteries all emphasize meditation. Religion is about non-local consciousness, and when you strip away the dogmas (I've written a series of papers on this), most religious rituals are designed to allow individuals in community to have non-local consciousness experiences. All religions begin with a non-local consciousness experience of a single individual. *Like Jesus or Buddha or Mohammad.*

Most people who get into this kind of research start from the premise, is this real? But by the time I had spent five years reading all of the Cayce readings, Steiner, Blavatsky, Gurdjieff, and the parapsychological literature, I didn't have any question about the reality of it. My question was, how did it work and what could you do with it that was useful? I began developing a protocol which eventually emerged into the Mobius Consensus Protocol that I used to remote view archaeological sites:

Cleopatra's palace, Mark Anthony's palace, the Lighthouse of Pharos––one of the seven wonders in the ancient world, remnants of Christopher Columbus' caravel, the *Brig Leander,* the Talking Idol of Ix Chel in Cozumel, etc. They were all basically projects to explore how non-local consciousness worked and could you do anything of practical utility with it. And the answer to those questions was yes.

I ran parallel explorations using standard electronic remote sensing and discovered that you could not find those same things using standard electronic survey techniques. I had the best people in the world try and they couldn't find them but the RVers could. I came across an interview that Max Planck gave to *The Observer* newspaper in England in 1931. When they asked him, "You and Einstein are the two most famous scientists in the world. What have you learned?" To their astonishment, he said, "What I have learned is that consciousness is causal and fundamental. You cannot get behind consciousness; consciousness is the source of all. Space-time arises from consciousnesses, not consciousness from space-time." In a speech he gave to a group of physicists in 1944, he said, "I have spent my entire life studying atoms and molecules and I'm here to tell you that they don't exist." *They are just potentiality?* They are manifestations of consciousness.

That led me to Einstein who described what we call reality as an optical delusion. *The Hindus call it Maya, illusion.* Yes, you discover the idea emerging over and over through history. The descriptive terms change; sometimes it is based on empirical observation or, more recently, on objectively verifiable measurement. It is all through both Eastern and Western religions as well as most of the founders of modern physics, the German school. We adopted their equations but not their conclusions. All of these people concluded that consciousness was fundamental, so I felt like I was on safe ground. I began developing protocols to test this idea of non-local perception and to think about what it really means when you say consciousness is causal, fundamental, interconnected and interdependent. We live in a matrix where everything is a manifestation of consciousness.

The whole Middle Bronze Age idea upon which the Abrahamic religions are based is that we have dominion over the earth, that humanity is a different order from anything else on the earth, that males should be dominant, and women subordinate. Those Abrahamic attitudes will not get us through climate change. *Jesus was a feminist.* Yes. Male dominance is not Jesus, that all arises from reflections of the Middle Bronze Age of the Middle East and had nothing to do with Jesus' teachings. Jesus is baptized by John and goes into the desert to meditate and awakens. Mohammad goes into the sacred cave Hira, where he meditates, the angel Gabriel appears to him and he awakens. Buddha goes to the hermitage of a meditation teacher, is taught to meditate and awakens. Note that the constant in all these stories is meditation, which is to say each founder mastered the discipline of attaining and sustaining IFA, which allowed them to open to non-local consciousness.

If the founder is charismatic enough that people listen to what they say and the collective consciousness of the community is attuned to what they are saying, then their teachings take root in that culture. If that resonance doesn't happen, they are just cranks. What we call genius is much the same. The Ah-ha! moment of insight is an individual experience. What makes it genius is social acceptance. Edison was not the first man to invent a light bulb since he was actually the 37th to get a patent on light bulbs. He was simply the man who did it at the right time in the right way and that makes a big difference. Darwin's grandfather Erasmus Darwin wrote a two-volume poetic text called *Zoonomia* in which he laid out a great deal of the principles of evolution, but nobody paid any attention; it was considered just an eccentric poem. When Leonardo was excavating the San Marco Canal in 1492 he came across fossils and figured out what they were, but nobody could hear it. The culture of the time thought that fossils were frozen fairies or that they were the result of seminal gusts from the gods.

"Religion" is an act of collective intention expressed through its scriptures and rituals. From a consciousness science view, the rituals are an empirical neurobiology process whose purpose is to allow some individuals in the community to have a non-local consciousness experience. We are now able to explore this scientifically. For instance, if you look at religious ceremonies, stripping away the dogma, what you get is a gathering. Does meeting at that place do something to the location that can be identified and objectively measured? Put another way, what makes it a sacred space? We know from research that IFA aimed at a particular location will make it more numinous. For instance, for RVers, it's easier to see Chartres Cathedral than it is to perceive a warehouse of the same size because Chartres Cathedral has been the focus of millions of acts of intentioned awareness often in a heightened emotional state, which adds informational enrichment. Consciousness manipulates information to produce reality.

So it creates a kind of thought form? Yes, thought form is a cultural term developed by the Theosophists. It is an informational--thought--structure just as it says. In trying to understand the non-local the great mystery question to me--for which I don't have an answer although I've been asking this question for going on 50 years--what is information? *Information is consciousness?* If you look at the experimental work, it shows us reality is a construct of ICA. It is both a physical space-time manifestation but also acts of intentioned observation informationally enriched, the information architecture; that's what Plato was trying to say about his idea of the ideal forms. Acts of intentioned observation add information to the non-local architecture; it's like turning on a neon sign. Jung called this information *numia*.

What about when photons or electrons are entangled and separated *and they have non-local effects on each other?* I don't think the evidence is at all conclusive about QM's role in the process. Whether non-local consciousness interacts with quantum processes is speculated upon, but the experimental evidence that can give us an unimpeachable understanding of that is not there. I am an experimentalist so I just care about what the experimental results are. *Information connects entangled particles?* The particles are an expression of consciousness, which was Planck's point. But exactly how that occurs in the non-local domain isn't clear. Quantum entanglement exists. What we don't know is how consciousness affects quantum entanglement. Dean Radin* has been doing a double-slit experiment trying to explore that idea. We do know that conscious intentions can change the molecular structure of substances like water. We know that conscious intention can affect the aesthetic appreciation of a substance like chocolate. *Like wine that you've done experiments with?* Yes. We know that conscious intention can have both a therapeutic and an anti-therapeutic effect on another organism. We can measure all this through experiments. We may not know at depth how it works, but we do know how to control variables that can directly affect what happens.

Like healing and voodoo? Those are empirical insights. The idea is that an individual can hold IFA about another organism and have an effect on its well-being. I did a series of experiments that showed when you focus intention on water you change its molecular structure in an objectively measurable way. *It bonds differently?* It changes the bonding relationship, yes. I did research in which I took the same bottle of wine, split it in two, had meditators focus on half of the bottle and then have people taste them, telling them it was two different kinds of wine and asking them which was better. Overwhelmingly they picked the wine that was the focus of the meditators who had focused on it before the tasting. *That was when you were in Virginia and Washington, DC, one of your earlier experiments.* Yes, and others have found the same.

Dean Radin did a series of experiments where he took tea and chocolate and split them in half, one of which was given to meditators. Gary Schwartz* did a study in which he had Hebrew words and fake Hebrew words and asked people questions about them. He found that the real Hebrew words always rated higher than the fake Hebrew words. Suitbert Ertel, a German researcher, did the same thing using Japanese characters in which some were real and some were not; people always chose the real words from the fake words. When he presented them upside down they got a lower score than when he presented them right side up. This tells us that it is possible to change the informational structure in the non-local domain. Religions are essentially empirically developed protocols, called rituals, designed to allow people to experience non-local consciousness.

Collective intention and awareness produces powerful social effects. We know that people have presentiments, precognitive awareness of things. I think the fact that we are becoming a majority-minority country and that climate change is about to severely disrupt human civilization has produced a fear fugue. This is further skewed because 27% of the population have overactive amygdalas, fight or flight, stimulated by fear, and correlate with conservative political and religious views.

To get back to your biography, how did you get from the East Coast to Los Angeles? I became the Specialist Assistant to the Chief of Naval Operations and lived through Watergate and knew a number of people involved. I went home one night and said to my wife, "I've got to get out, we've got to leave. I can't tell the good guys from the bad guys anymore and I can't operate like that." I resigned and decided to go to Arizona and write a book on the use of RVing in archeology. Just as I was finishing writing *The Secret Vaults of Time,* I visited my good friend Peter Tompkins who, along with Chris Bird, wrote *The Secret Life of Plants.* We talked about writing a biography of Manly Palmer Hall, whom neither of us knew. I went to Los Angeles to meet Manly and Henry Drake, his vice president, for whom I developed great respect. He made himself almost invisible at the Philosophical Research Society yet ran every aspect. It became clear right away that Manly didn't want a biography written about him. I went back to Arizona and went back to working on *The Secret Vaults*. Some weeks later they offered me the position of Senior Fellow and I accepted.

When I was living in Washington, a friend sent me the translations of papers written by Leonid Vasiliev, a Russian physiologist in St. Petersburg. Most people in the parapsychology at that time in the 50s through early 70s thought of the non-local as being like regular signals. *Electromagnetic.* Yes, Frederic Myers coined the term telepathy, for instance, in 1882, because the researchers at that time were beginning to look at consciousness again and many of them were also people like William Crookes who created the cathode ray tube. Many of them were involved with electromagnetics, the idea was that you were sending a signal and it was going mind-to-mind communication. In the early 1970s, a researcher in Canada, Michael Persinger, wrote a paper in which he argued that if psi existed, it was an electromagnetic phenomenon in the ELF (extremely-low-frequency radiation), 3 to 300 hertz range.

I didn't think a signal was the mechanism for something like telepathy because I could not imagine how you could locate a person. I did research on electromagnetic power long enough to learn that the more power you had, the stronger your signal was. So, the question was if it's the signal that you can send all over the world, where does the power come from? I did a lot of research and talked to people who were doing research about how much power the brain had and it actually has enough power to light up one of those little LED lamps, which is nowhere near enough power to send a signal any distance. Also, how could you deal with NDEs? I met George Ritchie in the 60s who was one of the first people to write about NDEs after he had one. If the brain was dead, where was the power coming from and how could you make the signal work?

Vasiliev was asked by the Communist Central Committee to look at whether this kind of human performance was electromagnetic. He was well-funded and meticulous. He eliminated the EM spectrum piece-by-piece by putting people in Faraday cages down into mine shafts or caves. He would ask them to perform non-local perception tasks at increasing distances. He finally came to the same conclusion as Michael Persinger. Vasiliev found that the only part of the EM spectrum remaining after he had shielded the participants from all other parts of the spectrum was ELF. He had eliminated everything else. He wanted to put a researcher and some participants in a submarine, because sea water at depth can shield all EM radiation, including ELF. He went to Admiral Sergey Gorshkov, who was the father of the Soviet Blue Water navy, but he wouldn't do it. When I read that I thought that would be the way to do the experiment.

In Los Angeles, I stayed with Don Keach, a Navy captain and friend who had just retired. Along with Don Walsh, a commander who had made the deepest dive in history, they took over the Institute for Marine and Coastal Studies. He told me, "You know that crazy experiment you want to do. Well, we've got a research submarine coming down for the summer and we'll pay for three days for you to do your experiment." It was a gift that changed my life that came to be known as "Deep Quest." I filmed it with Leonard Nimoy as the host. You can see the archaeology part of it on YouTube.[1]

The first task was to use RVing to locate on a sea chart a previously unknown wreck on the sea floor. I focused on archeology because it offered very meticulous, triple-blind experiments. Everybody agrees they don't know the location or have a detailed description of what would be found there. If we were successful it would show that sea water was no barrier to non-local perception, as it is to EM. Second, I already knew it was very hard for RVers to get analytical information and that you didn't want to ask them to do analytical cognitive tasks. They can describe the shape, the color, how many parts it's got, what it feels like, what it smells like, but when you ask them to name it you are asking them to analyze. But what if you needed to get analytical information conveyed? I had an idea I wanted a test that would address that limitation and also answer the other side of the EM issue. Could viewers when submerged acquire information from the surface?

I got the idea while researching a speech about law of the sea for Elmo Zumwalt, then Chief of Naval Operations. As I looked through the naval archives, I ran across a picture of a codebook created by Admiral Lord Nelson for the Battle of Aboukir Bay in Egypt in 1798 in which Nelson defeated Napoleon's admiral. He did that by creating a codebook in which colored flags were actually commands. I thought I could do the same thing by associating an object of location, which I knew RVers could reliably do, with analytical terms like names or numbers. In Deep Quest I put people in the submarine and asked them to describe where people were hiding and associated the hiding place with a message. So if you are hiding under the oak tree, that means go to the supermarket. If you are hiding under the overpass of the street, that's to go to the gas station. I called it Associated Remote Viewing or ARV for short.

How did you meet the SRI researchers and how did they get involved with Deep Quest? I met Ingo Swann a few weeks earlier, and he had done the map phase, so I asked him if he would do the submerged viewing and he agreed. He suggested Hella Hammid who was also working for SRI. I already knew Ed May, a nuclear physicist who was working in the SRI program, and I knew he, Hella, and Ingo had worked together. So I asked Ed to do the interviews. That's how SRI (Stanford Research Institute) got involved. And every aspect of the experiment worked.

How did ARV become involved with financial investment schemes? About a week after we had done Deep Quest, I went to a dinner party where a skeptical surgeon didn't believe my story. He said, "The only thing that would make me believe it is if you could win a horse race because they're random and you can't predict how they are going to come out." The task for the viewers was to describe a place they would be taken two days in the future at 4:30 in the afternoon. There were six horses, thus, six locations. Triple blind. I didn't tell them that the viewing would predict the outcome of a horserace—only that they would be taken to a location. They both described a place associated with the sixth horse in the sixth race.

The next night we all went to the races along with my eight-year-old daughter, and we bet $2.00 on the sixth horse in the sixth race. We won $14.00 and were jumping up and down screaming and yelling. I wrote a paper saying you could use this to predict things like the stock market. To test the idea we began an ARV investment program using the Standard & Poor 500. Russ Targ* started doing Silver Futures and made over $200,000. Harold Puthoff ran another ARV investment which earned $26,000 that he needed to start a Waldorf school. Mobius did 42 weeks on Thursdays calling the S&P 500 for the next day. Did it go up, down, or stay the same? We started with $5,000 and made $150,000. But, after 42 weeks I realized I wasn't getting any other work done; everyone was focused on are we going to win this week. The most sophisticated ARV ever conducted, for which I was a viewer, was done by mathematician James Spottiswoode who had worked with both Mobius and SRI. He developed a complex redundancy protocol to determine the six numbers in the California lottery and picked the winning sequence but was unable to get the tickets printed out in time for the call.

In my view, the way to use the ARV protocol is adjunctively when in a general consensus protocol you need to get some highly analytical information. RVing is a technology that you can use to solve primary tasks, but it's not a get-rich-quick scheme or a silver bullet. Basically it's the same thing as using side-scan sonar, or ground penetrating radar. It's a sensing technology; it's not a searching technology. You just go to the place where it tells you to go and it's either there or it's not. By the late 70s, early 80s, big labs with a serious RVing research program included Mobius in LA, SRI in the San Francisco Bay Area, and PEAR, Princeton Engineering Anomalies Research group in New Jersey.

After we did Deep Quest, two women historians asked me, "Could you use RVing to find the Tomb of Alexander, the Lighthouse of Pharos, Cleopatra's Palace and Mark Anthony's Palace?" That began what came to be known as the *Alexandria Project* where we found those sites. We also found what I believe is Alexander's tomb and I believe we know where his bones are buried but I can't get the monks who have the bones to do the DNA test. The site that the RVers picked was underneath the Mosque of Nabi Daniel in Alexandria, Egypt. In 1979 we were digging under that mosque and everything that the RVers were telling us, we found. There were 11 viewers, two of them on site. You give the viewers a map or a chart and ask each of them to locate something on it. Each gets their own copy and then you see where the choices overlap, to create a consensus. In this case, nine of 11 viewers had picked the same place underneath this mosque. My first question was, is the tomb of Alexander the Great within the area represented by this chart? Second question, if it is, please locate it as specifically as possible.

I took George McMullen and Hella Hamid to the edge of the circle of the consensus zone, and said, "You are now at the consensus zone that you selected along with the others. Here's a wooden stake. Please walk around this consensus zone and locate very specifically where it is." I went to the director of mosques in Egypt and asked for permission to dig, which we got, until the incident in the American Embassy in Tehran halted the dig. McMullen said, "Don't be depressed, Stephan. The bones are not there anyway." He explained that in 641 CE, a group of monks came down from their monastery in the desert, masqueraded as merchants and went to what was then a church, not a mosque. They went to the ossuary, the place where they kept bones, and they gathered them because they believed some of the bones were John the Baptist, and they took them up to the monastery in the desert. *He was RVing all that?* Yes.

I asked him how I would know whose bones were whose. He said, "When Mark Anthony came to Alexandria with Cleopatra, he went into the Soma (that's what the tomb was called) and he had them open the coffin. He took out his red Roman war cape and put it on the body as a way of showing that he was the heir to Alexander. After the Christians took over the city, they didn't care about the Tomb of Alexander because he wasn't Christian. Water leached into the coffin and seeped the red dye from the war cape into the bones."

A couple of weeks later, I get a call from NBC TV news saying, "You are the only archeologically-trained film crew in Egypt. There is a community of monks in a monastery up in the desert toward Cairo, and they think they found the bones of John the Baptist. Would you go up and film it for us?" We went up to the Monastery of Saint Macarius in Wadi El Natrun, an oasis. I went up and knocked on the one little door, and the Prior, Father John, opened the door. He agreed to let us film where the bones had been found. When I asked him where the bones came from, he said they came from underneath the mosque. He said that in 641 CE, when the city was turned over to the Muslims, a group of Christian brothers masquerading as merchants went down and gathered up the bones and put them in leather bags tied on the back of donkeys—exactly what George said. It was astonishing. I asked if there was anything unusual about the bones. He told me a few of the bone were stained red, but they didn't know why. There was no DNA scan at that time but it seemed a very compelling coincidence.

A number of years later when the DNA technology had been developed, and they had found the Tomb of Philip, Alexander's father, his DNA was tested and published, and things changed. Were these bones Alexander's? There was now a way to answer that question. I have been asking for some years for the monks to let me come and do a DNA test on the red-stained bones. The problem is that this is a Coptic Christian monastery and they don't want a lot of attention, fearful it will result in persecution, given present day Egypt. Also they think some of the bones belong to John the Baptist and are holy. *You wrote several papers on this work, and Deep Quest before it. Are they available?* Yes. You can go to Academia.edu or ResearchGate, search on my name and all the research papers are freely available. *The Alexandria Project* is the book's title, the *Deep Quest* and *Alexandria Project* films are on YouTube, filmed as the events happened.

Do you think Atlantis and Lemuria and those kind of supposed ancient civilizations existed? Have any of your RVers looked at them? I got very involved with Atlantis. Colonel Sanders, who created Kentucky Fried Chicken, had a daughter named Margaret Adams. She and I and some other people started the Marine Archaeological Research Society to find Atlantis. But as mitochondrial DNA research developed it became clear to me that there was no Atlantis or Lemuria because if there had been, it would show up in the DNA and it doesn't. I've gone from being a strong proponent of Atlantis to believing that it is a myth and possibly a precognition of our future. *Edgar Cayce spoke at length about Atlantis.* I think he was wrong and was also wrong about Egypt. He said very clearly that the beliefs and expectations of the questioner had a strong influence on how he did his search, like a Google search. In non-local consciousness, a strongly held belief is as real as an actual thing and everyone involved in the shared intention are players. All the people who were asking him about Atlantis were fanatics on the subject. When they asked them about it, their intention shaped his search in the information matrix where the myths about Atlantis existed. He did the same thing with ancient Egypt, so all the Egyptian material like the pyramid was built 10,583 years ago, none of that is correct.

What led to your book Opening to the Infinite? I kept getting asked how to do RVing and I was invited to do workshops on it. I listened to the questions people asked and decided to write a book presenting everything science knows about how to do RVing. It teaches you how you do it, what you can do with it, and what it's telling you about what you are as a human being. RVing is a bell curve with a few people at one end who are exceptionally good at it. You can't train for it in the sense of learning a formula, but you can get better at it. The key to RVing is the ability to attain and sustain IFA and meditate. In *Opening to the Infinite,* I teach a way of meditating designed specifically for people with modern minds, because most meditation techniques developed centuries ago when people lived in a slower and quieter culture. I created a technique called *Meditation for Modern Minds;* you can go to my website and download it. Meditation is important for more than RVing. We know, for instance, 11 hours cumulatively causes improvement in the prefrontal cortex of your brain and causes your stress and anxiety to go down. Thousands of papers relate the benefits of meditation: you sleep better, you're more creative, your IQ goes up, you're less stressed, your sex life gets better and your relationships with others get better. I tell people, "If you really want to change your life, develop the daily practice of meditation."

How did you get from running an experimental laboratory to the social research you do today? I am still an experimentalist. I am doing a large RVing study. The research taught me that all consciousness is interdependent and interconnected, and that we are looking at an information phenomenon. From that perspective culture is created by a consensus of intention; that's how the individual becomes social. Also I had done a project RVing the future and became convinced humanity faces a civilization-threatening crisis with climate change. The only way through this is by integrating consciousness into science, technology, and governmental policies. Things I learned about individual access to the non-local had social consequences.

In The 8 Laws of Change, *you include a chapter on how to meditate. Is that what you mean by extending from the individual to the social?* I developed the "theorem of well-being." It says any government policy whose principal function is to foster well-being will be more efficient, more productive, easier to implement, nicer to live under, and much cheaper. You hear people ask, for instance, how would we pay for universal healthcare? They don't seem to understand that we spend more on health care by a very large measure than any other country in the world. *With worse outcomes.* Yes, we are ranked 37th by the World Health Organization. If we could do what Norway does and get to the 11th position and spend 7.6% of our GDP on healthcare as they do, we would free up about $1.3 trillion a year.

In the studies of happiness and of well-being, the Scandinavians come out at the top. How is it that they can do it and we can't? Because we are not trying to do it. The social outcome data is clear. Just start with literacy, education outcomes and obesity. American children are not a priority, despite what we tell ourselves. In this country rugged individualism is prized and promoted over the well-being of the country as a whole; we do not recognize that in the non-local domain consciousness is an interconnected functioning whole.

In The 8 Laws of Change, *you mentioned schools in crisis that taught meditation and built time into the daily schedule produced much better outcomes for these students.* San Francisco took the four worst schools in the school system and developed what they call a mindfulness period, a kind of meditation period. The violence in the school dropped, along with the dropout rate and the graduation rate went up. When they tried to stop the daily meditation, the students rebelled and demanded that it continue.

What about the Transcendental Meditation experiments where they meditated and the crime rate went down? Conceptually it's a very engaging idea. I spent three years and a great deal of money trying to replicate that effect under conditions rigorous enough to make the conclusion plausible. We're talking here about a series of experiments that were done by Maharishi followers back in the early 80s, in Atlanta. Unfortunately there were a number of uncontrolled variables that could have produced the result. With the help of the Los Angeles Police Department data, I designed a study which addressed those variables. I got a church to agree to let the meditators gather there. We were going to put them in as the cleaning crew so nobody would notice them. I went through police department data and, as I worked with their voluminous and very detailed statistics, I realized I could see how the Maharishi experiments were flawed. For instance, the police taught me their research showed if just one police car drives through a neighborhood, crime will go down. Conversely, if a car is parked on a street and the window gets broken and isn't repaired within two weeks, crime will go up. The Maharishi experiments did not control for any of those variables. If you can't control for such variables you can get false positive or false negative results and be entirely misled.

How have you been involved in social movements you write about? Four times in my life I have been involved with changing history and I've paid attention to how that happened. When I began doing this research, over 30 years ago, I thought I understood how social transformation worked. As I got further and further into it I realized nothing I thought was important. Usually the discussion is how do we get enough money to do this, how do we get the auditorium to hold the conference. In fact, success is the result of something much subtler, the eight laws. Using them is how Gandhi got independence for India and how Martin Luther King got the civil rights movement to work.

You've been involved in the civil rights movement and the consciousness movement. What were the other two? In the 50s and 60s it was civil rights. Much of the 70s I spent as part of the small group that transformed the American military from the elitist conscription organization of the Vietnam era to the all-volunteer meritocracy that we have today. The 80s were focused on citizen diplomacy, the idea that it was possible for ordinary people to create meaningful channels between supposed enemies, in this case, the Soviet Union ecology movement. (For more about the 8 laws see the book's blog.[2])

You've written five non-fiction books that have won awards. Why did you also start writing novels? I realized Millennials and Gen Zs aren't reading a lot of non-fiction. They mostly read novels, so I decided I could put the same information in the novels that I put in the papers and continued to write papers and non-fiction books. My novels are *Awakening,* the first one, a novel of aliens and consciousness. *Do you think ETs exist?* Yes, but I believe almost everything we think about aliens is wrong. If they had invasion in mind, they would have done it centuries ago. In order to develop space travel, you have to develop the idea that consciousness is fundamental, which is what I think aliens are really all about. I believe they are monitoring whether we are going to get through this phase or not. They may be helping us. The second novel is the first of the Michael Gillespie Mysteries series, called *The Vision,* a novel of time and consciousness. The third novel is *The Amish Girl,* the second volume of the Mystery series, a novel of death and consciousness. It's based on a true story, which *Nova* included in a TV program about the disappearance of a young Amish girl. The RVers found her, and accurately described what had happened.

Are you optimistic or pessimistic about our future? The question of how we respond to climate change is going to determine whether we are going to be destroyed or go forward. I'd say at this point, it's about a 50-50 toss-up. Are we going to revert to primitivism? Are we going to simply vanish as a species like the dinosaurs? We're going to have hundreds of millions of climate refugees for instance, we're going to see massive disruption of cities. In the US we're going to have three major migrations: away from the coast because of sea rise, out of the southwest because of lack of water and temperature increase, and out of the central states because of violent climatic events like tornadoes. I'm not sure the United States is going to survive: I'd give it about a 60/40 probability, 60 being it won't survive.

What is your current research focus? I am currently doing a large population RVing project to look at the year 2060; supported by the BIAL Foundation and Atlantic University. I did an earlier similar project. From 1978 to 1996, I had 4,000 people RV the year 2050. Everything they told me would happen between now and 2050 has happened or is happening, although I found much of it unbelievable at the time. I started the 2050 project because when I left government in early 1976, like most of the people in the geopolitical community, my central concern was nuclear war. Yet, from the beginning, the viewers all said there would not be nuclear exchange between the US and the USSR. I commented that the world must be much safer but the RVers said, "No, it's much more dangerous," because of terrorism. In 1978 terrorism as a massive threat didn't exist. I asked what happens with the Soviet Union, they said, "I don't know but it doesn't exist anymore." Now we know it collapsed in 1991.

About healthcare, the RVers said there is going to be a series of epidemics, starting with a blood disease that crosses over from primates to humans in Africa and is brought into the US. It kills millions of people worldwide. In 1979 I went to a friend who was a senior official at the National Institutes of Health and asked if he knew of such a blood disease. He didn't but in 1981 the first AIDS cases began to emerge and 35 million people have died of AIDS since then. The RVers also predicted the collapse of antibiotic medicine. They said climate change was another really big issue.

When I visited Virginia Beach, I did a series of sessions to ask what Virginia Beach was going to be like in 2050. They said much of it will be underwater because the sea level had risen so much. I didn't know anything about climate change until 1991 when I read a paper in the journal *American Scientist* talking about ice coring. In Phoenix they predicted there are few people in this city because it's too hot and there's no water. The Colorado River is not going to be able to provide water and the lakes like Lake Mead are drying up. It's important to remember that none of this is predestined. When you ask people to do precognitive tasks, you're asking them to read the highest probability of that target future at the moment. That is why I am now doing 2060. I want to see the difference between 2050 and 2060. I think my next book will be non-fiction: *Sacred Mysteries: Science, Religion, and Consciousness,* followed by *the 2060* findings, then a novel.

What do you do for fun while doing all this research and writing? I spend time whenever I can in wilderness. Throughout my life, since I was a boy, I've done hiking, backpacking, canoeing, scuba diving, anything to do with boats. I live on a rural island in the Pacific Northwest and along with my wife I hike the trails in the forest around our property almost every day.

Books

Nonfiction

The Secret Vaults of Time, 1978
The Alexandria Project, 1984
Mind Rover, 1989
Opening to the Infinite, 2007
The 8 Laws of Change, 2015

Fiction

Awakening: A Novel of Aliens and Consciousness, 2016
The Vision: A Novel of Time and Consciousness, 2017
The Amish Girl: A Novel of Death and Consciousness, 2019

Papers

www.academia.edu
www.researchgate.net
https://stephanaschwartz.com/biography/

Endnotes

1 https://www.youtube.com/watch?v=NKMN509cOJQ
2 https://visionaryscientists.home.blog/2019/07/31/how-to-create-social-change-stephan-schwartz-interview-on-his-book-the-8-laws-of-change/

Patrizio Tressoldi, Ph.D.

Science of Consciousness

Photo by the author

Questions to Ponder

What do ancient Indian religions and Quantum Physics have in common?

What are the results of Dr. Tressoldi's experiments where intention changes matter? What are the practical applications of the Mind Switch, Implicit Telepathy, precognition for drivers, etc.?

What is the therapeutic application of experiencing past-life deaths?

What else might exist besides space-time?

I was born in Italy, in a small city near Venice, 31st March in 1953. *You are a Pisces.* In my family, my father was the main worker, and my mother was a housewife. I was the first of the three boys but the smallest one because the other two are taller than me. My parents insisted on the importance of study: This was the big message that we got from my parents. My two brothers are not involved in scientific investigation. They're normal people with different working interests.

Were you raised as a Catholic? Oh yes, in Italy we have a monolithic religion and I was grown in a typical Catholic cultural environment where I practiced lessons from the priest and was baptized. After my adolescence, I clashed against the dogmas and tried to find answers to my spiritual questions by searching outside the Catholic doctrine. *Hindu Vedanta and Buddhist traditions viewed reality as illusion. Those sages knew thousands of years ago what quantum physics leads us to now.* We may say they are the best psychologists because all their writings are about who we are and in particular the function of our minds. What astonished me as a psychologist is what they taught 2,500 years ago is now verified by scientific investigations. *They came across it from interior introspection rather than external observation so their process was different but it worked.* This is a very radical revolution to find answers to our scientific questions searching inside us and not outside. It's clearly a revolutionary concept that changes the dominant paradigm. Currently the studies about meditation, mindfulness and yoga are mainstream in science. Of course, some people are only interested in behavioral effects or in the correlation with brain activity, doing normal science.

Does that link to your advocacy to honor first-person experiences? The science is very clear. You cannot investigate pain, pleasure, and emotion by simply looking from outside the person. Even measuring their physiological reactions is not sufficient. Of course there are people who cannot express their feelings, their emotions, such as people with severe brain injuries. But in general if you want to know their pain level, you must ask them in order to grasp the phenomenological experience. *It's simple to say, on a scale of 1 to 10, what's the pain in your knee today?*

When you went to university, as an undergraduate, you studied psychology? Yes, my scientific curriculum was always in psychology to satisfy my curiosity. I am curious about physics, about education, about clinical psychology, but my career was in psychology. *You've written about dyslexia in students.* The first part of my academic life was dedicated to learning disabilities in children. But underlying that I always had a big curiosity about what is reality, who am I, and so on. In the last 10 years I abandoned my first interest in learning disabilities and clinical psychology to devote my main interest to the potentialities of our mind.

What have been your most difficult challenges? Honestly, I haven't had very severe problems regarding my health, my family, or my professional activities. In my experience in the academic environment, I realized that scientists are humans. I expected scientists in general to be more open to new discoveries, to breaking the wall of old knowledge but it is not so and that makes me a little sad. *What helped me in my university career was Frans de Waal's book* Chimpanzee Politics, *about the chimp dominance hierarchies where the alphas try to look bigger, do favors to build coalitions, etc. It helped me to know that we're primates.*

The University of Padua has been supportive of your work, I assume. For most of the academics I've interviewed, certainly in the US, not so much in the UK, there's a feeling of if you're going to do this work, keep it to yourself. Yes, this is a big problem for my investigations, but I was respected for my scientific methodological skills and for doing good research according to the best standard. My colleagues respect me as a good scientist, as documented by my CV. When I declared my interest in fringe aspects of my investigations, most of them were curious. They don't want to be involved with us directly but often in my office, they ask, "There's news about your investigation. Can you tell me something about it?" Outside of my office they don't ask. But many people in my department––one of the biggest psychology departments in Italy--know very well what I'm interested in.

They are waiting for breaking the wall of the dominant paradigm that officially negates the possibility that the mind can acquire information instantly or can influence electronic devices like RNGs. This is what happened for the study of meditation. Before the 1980s, very few people investigated meditation, now referred to as "contemplative discipline." After the 80s there was a dramatic increase and now investigating meditation, mindfulness, yoga and contemplative discipline is mainstream. I am optimistic that in less than 10 years what we are investigating now will be mainstream. *Do you see any difference in young faculty and young graduate students?* Yes, many of them ask me information about my field of investigation because they don't study the potentiality of the mind. They study classical cognitive psychology, neural correlates of mind functions and so on, but nothing about the power of the mind as evidenced by the placebo effect, for example, so they are very curious.

What about your own experiences of meditation, precognition or synchronicity?

I declare myself a normal human without particular exceptional experiences, but I try to educate my mind. I'm a long-time meditator. I developed my PK skills a little bit more. We are doing very interesting experiments and I'm one of the participants. We try to influence electronic devices, in particular RNGs. Recently we are trying to influence the memories of digital professional cameras which cost around $2,000, which are very sensitive to light. We try to generate images on the memories of the digital camera with the shutter closed in a dark environment. If we are able to print even simple images such as a geometric shape intentionally, it will be a great demonstration of the potentialities of our mind. We are all skilled participants and meditators in our lab.

You mentioned in an article that photons may go from the mind to the object and that's what's changing the random zeros and ones from *the RNGs.* We published a paper demonstrating the possibility of influencing detectors of photons from a distance.[1] Working with the Rhine Center in the USA, more than 7,000 kilometers away from Italy, we found a clear demonstration that mentally we increased the number of photons emitted in their dark room. John Kruth* simply acted as an activator of all the apparatus and didn't know when we tried to influence it. He sent us the data and we saw that they changed depending on our intention to increase the number of photons in the room.

The PEAR studies at Princeton found the strongest effect on the RNGS was with two people who are bonded. Do you find that one person is not as strong as a group sending intention? Yes, a group is always better than a single participant unless the single participant is a superstar. The experiment with the Rhine Center was conducted with about four participants in a working group similar to the psi-photo experiment.

Do photons come from your mind to the Rhine Center? This is not something that arises from our brain to pass through walls. A friend who is a quantum physicist said it's impossible to transmit photons from your mind without the photons being absorbed by obstacles. We sent the intention of photons directly to the apparatus. We don't send the photons; it's an intention that becomes concrete. I send information so there is nothing that travels from Italy to North Carolina. That's a clear and astonishing hypothesis that your purely mental intention creates reality. Some quantum theorists postulate that we can create mental entanglement, but it's important to distinguish mental entanglement from physical entanglement, where particles of physical stuff interact directly and then can be separated and influenced at a distance.

Some people use the word "information." It's important to not confuse physical with mental information. Many say that matter is nothing more than information. *Consciousness?* Probably they are two faces of the same stuff; one is mental, the other is physical.

Our main aim is to demonstrate the capacity, the potentiality of our mind by using very cheap devices that can be used by normal people. *Like the plasma ball used in IONS experiments.* I am convinced that resistance to psi phenomena will decrease dramatically if you demonstrate it with practical applications to encourage revising the theoretical interpretations of mind and reality.

We may be releasing photons but they don't travel to what we're trying to change. It may be that they send out an informational signal? For an intention to become physical and materialize seems strange, but the placebo effect demonstrates this principle. For example, what the physician suggests may decrease my pain, especially if she wears a white coat. Your body transforms this intention into molecules for decreasing your pain. If this is not a miracle, I don't know how to name it. The investigators of the placebo effect are observing that the intention activates the same physical substrates that are triggered by the normal anti-pain pills. Decreasing the pain was created magically from the intention of the patient and physician. What we are trying to demonstrate is that it's possible to materialize the mental intention outside our mind without space constraints. It's only a difference of distance. *It can go the other way too, so if I believe someone is sending me a curse, I could get sick due to my belief.* Absolutely, we must be very careful about what we allow to pass in our mind. This is the reason why good feelings are recommended for our physical health.

I read that one happy person affects people that they're not even closely connected with in a kind of chain.[2] *I imagine it would follow for people who are angry too.* Researchers at Maharishi University recruited over a thousand meditators simultaneously in order to decrease negative effects in the environment, such as to reduce crime. This is a clear demonstration that collective mental intention can influence the quality of life of many people around us even if they don't know that there are people trying to positively influence their life.

You've written a book in Italian titled La Parasicologia. *Do you have another book that you're thinking about or are you busy with the over 90 papers that you've written?* The scientific dissemination is an important topic but also very important for non-academics. We have two Facebook groups: one dedicated to psi experiences and the other to the science of consciousness; they are another way to disseminate with plain language what we are studying.[3] Dean Radin's books are among the more interesting, and another very good writer is Deepak Chopra so it is not necessary for me to write a book in English.

You looked at mind control of RNGs from a distance finding weak effects that are very statistically significant. Why is it a weak effect? It depends, but we recently demonstrated that even a weak effect can influence the output of the RNG. We wrote a recent paper about the Mind Switch experiment, a very cheap device that costs less than $100. We have posted all the information for building this apparatus and we offer the software to analyze the data.[4] We demonstrate it is possible to switch on an apparatus based on a true RNG. Even if the effect is weak, it's simply a matter of improving the software for analysis of the data. This is more effective and will be sufficient for practical application.

What does the Mind Switch do? How does it interface with the human and the RNG? With the apparatus, you can intentionally switch appliances on and off from a distance. You could have this kind of apparatus connected with your heater and switch it on with your thought. You can have an app in your smartphone to switch on or off every device in your house. *I'm coming home from work and want my air conditioner on. So, I think, "Device, turn on the air conditioner now," and it will be on when I get home.* Absolutely. Or you can send a sort of e-mail or SMS from your phone powered by a simple battery so if the Internet network crashed you have autonomy. That's an amazing finding. We have a proof of concept that it's possible, and it's shown in a YouTube video.[5] I think that if some start-up wants to invest in these kinds of devices, it would be a big investment. The problem is that many people are very skeptical about the possibility that the mind can influence electronic devices at a distance but it's not hard to make this apparatus; it's very efficient. It's only a matter of increasing the signal-to-noise relationship. It's simply a matter of software, nothing particularly sophisticated.

We did another experiment trying to send an audio-visual signal from one mind to another mind by analyzing the information.[6] There is a clear demonstration that you can send at least two different signals from one mind to another mind at any distance. The method is simply a matter of an algorithm for analyzing the signal in the noise of the EEG activity sent from the transmitter. Intentions can be transmitted and detected physically through mental telecommunication. I expect some courageous start-up to invest their money and skills to exploit these mental technologies, which are ready to be improved upon and put on the market.

MIT is working on enabling a person with a missing leg to think "walk" and machinery enables them to walk. In this case, it's simply a translation of your EEG activity for external devices, related to normal physical transmission signals. We are trying to convince people that we can send mental signals that are not constrained by the limits of electromagnetic waves. There are no walls or obstacles that can alter or block this transmission. It's similar to the remote viewing where there are no limits to the possibilities to obtain information.

You did a study where two friends were connected to EEGs in different rooms. When one was stimulated with red lights or sound, their friend's body responded as measured by the EEG, right?[7] Exactly. This is what we call implicit telepathy because we don't require the receiver to describe what the sender sent to him or her because the information can be detected by analyzing it with sophisticated algorithms. At present we have demonstrated it's possible to send signals with at least two kinds of information. If we can improve this kind of technology, we can send much more sophisticated information. I am quite sure that DARPA [*the US Department of Defense's Defense Advanced Research Projects Agency that develops new military technology*] is doing a major investigation about this possibility for military purposes, but I don't want to offer this kind of technology only for military purposes. I want it to improve our quality of life.

I'm sure you know Bill Bengston's work where he's trying to capture a signal to do healing from a distance.* Absolutely. In fact, there are applications for health. Bengston is doing very advanced studies on the possibility to heal at a distance. It is another incredible opportunity for increasing our health technology based on mental ability. *Probably for as long as we've been human, people pray or send good intentions to others, so this is not a new technology. People knew that they could use their intention to help someone.* Yes, intercessory prayer is an ancient aspect of human existence. In this case, we refer to an external power or divinity that will receive our intention and transform them if he/she is in a good mood. In Bengston's case, there is a direct connection from an intention to the target, without denying that probably there are higher spiritual powers somewhere. But in this case, we don't need any contribution from others. That's the difference.

What did you learn from the people who were hypnotized and channeled descriptions of past lives, including their death experiences? We are investigating three anomalies or non-ordinary states of consciousness: OBE,[8] channeling[9] and past-life regressions.[10] The experience of death in the past-life regressions looks quite similar to the NDE experiences. We use hypnotic suggestion to obtain an intentional OBE and channel experiences and interview the participant. OBEs usually are not controlled as they usually are experienced after traumatic experiences, but our use of hypnosis makes this possible. We collected a lot of data about these experiences, including past-life recollection.[11] In general, the experiences are very positive and are transformative about our conception of death. They reduce the fear of death, increase confidence in an afterlife--very transformative and spiritual experiences. *Did you go through hypnosis to your past life?* No, because the hypnotists say I am not a good subject. *It's hard if you are analyzing the whole time, which interferes with letting go.* The characteristics of the participants are willingness, curiosity, and a good level of hypnotizability.

Michael Newton's books describe how he hypnotized people and focused on their experiences in between lives in the Journey of Souls *book series.* Yes, if we find similarities this opens up our curiosity about the question--is our life limited to our physical body or does it continue after death? *Is it safe to say after doing all this work, that you think some part of you will continue after death?* I am quite convinced of this, but I want to be a scientist. The evidence we have now is more supportive of an afterlife. If you are honest you must take into account all the evidence derived from investigation of reincarnation and our channeled experiences. It's too simplistic to say they are all fantasies of participants. Brain scientists have no hint of how to explain such vivid, complete stories.

To me the most convincing is Ian Stevenson and Ed Kelly's studies of children's memory of past lives. Over one-third of the children had birthmarks that were representative of how they died--that's improbable. Yes the evidence for reincarnation, mediumship, NDEs, and OBEs, etc., leads us to question where they come from. The problem is that many people are strongly committed to the idea that we are our body, and when your body ends, there is nothing to be investigated, although scientific investigation tells other stories we must take into account.

Have you done research with mediums or are you leaving that to Julie Beischel and her partner at the Windbridge Research Center and Gary Schwartz* at the University of Arizona?* Not directly, but we are doing a meta-analysis of all studies done since

2000. The evidence is based on certified mediums selected for their reliability and ability to produce information that can't be explained by something like cold reading [*a skilled person interprets body language and other cues*]. The evidence is quite convincing. *You also did a meta-analysis of presentiment or what Julia Mossbridge calls "predictive anticipatory activity."* Our lab investigations show that our body "knows" seconds before; our body reacts unconsciously. It is a paradox because we say that our body knows what is unpredictable.

A group in Canada is trying to use these kinds of signals for preventing car crashes so you can, for example, slow down your car and avoid a crash and save your life. We wrote a scientific paper about "Driving with Intuition."[12] We can use our skin conductance levels and many other psychophysiological indices. *So, your smartphone could be reading those measures while you're driving and a voice would say, "Slow down, you're headed towards an accident; we see your heart rate variability has changed."* Yes, it defies the laws of causality because you cannot have a reaction before something that you don't know. The question is, how is our body able to detect such things?

What it means is time isn't linear. We live in a dimension with linear time from past to present to future, in contrast to a dimension where there is no time and no space, where there is only the now. If we are able to shift between these two levels, we can discover and exploit all our potentialities. In fact many meditation traditions aim to center us in this particular state of consciousness where there is no time in that space. We can do different things than when we are in a three-dimensional space with a linear time arrow. *Do you think that the mainstream quantum physicists would agree that there is a dimension outside of space-time?* The field of quantum field theory is studying how matter can arise from the void. This may be the same phenomenon that we are experiencing in meditation, etc. Despite many interesting similarities I don't want to say they are the same. Probably they are the same at different levels of reality, one mental, one physical, while some say they are simply two aspects of the same reality: one mental, one physical.

Please define consciousness, the difficult question. Without consciousness, you can say nothing about reality. It's awareness. One debate is if reality exists without our reference, but something external to us probably does exist. *It exists as probabilities?* Without an observer, there is no reality. I discussed this with a physicist who said, "No, it's a measurement," not observation. *Isn't potentiality something?* Probably. Physicists push us to say that what we observe is simply a mental construct. There are many experts stating that reality is something internal to us and that there is no reality outside us. *Deepak Chopra spoke at the IONs conference in 2019 and concluded it's all an illusion.* I am quite close to his interpretation of reality. *It comes back to Vedanta sages who've said that for centuries.* Fortunately, science and spirituality are getting closer. In this decade many investigations about spirituality use scientific methods. This is very important for the future of humanity.

People who have NDEs like Eben Alexander totally change their view of spirituality.* I am very interested in direct experience, called paranormal or exceptional experiences, that cause you to revise your idea of who you are and what is reality. The famous Italian-American inventor Federico Faggin is the father of modern computer technology, the first commercial microprocessor. He changed his ideas after an OBE where he realized there is something more than the brain. *Susan Blackmore* had a major OBE and her conclusion after many decades was that it was just a function of a right temporal parietal junction in the brain that has to do with our sense of location in space.* Scientists can simulate something similar to an OBE but nothing similar to a real OBE. Brain scientists can't explain even normal experience. Every emotion has its typical brain marker but it's not sufficient to explain your phenomenological experience. At present there are no clues how it's possible to translate physical electromagnetic activity into a phenomenological experience. There are no clues at all. In fact, there are legitimate explanations that consciousness is something beyond the brain. *It's the Mind.* It's proto-consciousness that interacts with the brain and allows us to have such experiences.

If you are honest, a lot of experiences are more plausible with a proto-consciousness, an original consciousness that interacts with the brain to have a normal experience. But academics are very reluctant to accept that mental consciousness interacts with the brain despite the evidence that we are obtaining with our investigation about mental interaction from a distance. Donald Hoffman (at the University of California at Irvine, Department of Cognitive Sciences) offers a very strong empirical argument for supporting the idea that there is nothing outside that can't be explained by our inside mental activities.

Critics charge that parapsychology is weak because it's not easily replicable. In Etzel Cardeña's article on the experimental evidence for parapsychological phenomena, he counters that replicability is a problem for all social science.[13] There is a crisis of reproducibility in most sciences, including experimental physics, medicine, psychology, ecology, and economics. It's the cornerstone of scientific investigation to be able to duplicate research. Some experiments are more reproducible than others, as in the Ganzfeld clairvoyance studies. We accumulate evidence through meta-analysis. With a colleague in Australia, Lance Storm, we have done meta-analyses of experiments from 1985 to 2018. If you analyze all the scientific literature, the evidence is very clear; Ganzfeld is a phenomenon with repeated effects. The other strong phenomenon is remote viewing, always with strong individual differences. We are studying Rupert Sheldrake's work with telephone telepathy as many people experience knowing someone is going to phone, with a system for investigating this phenomenon. [*See our interview.*[14]] We are recruiting normal people but looking for an emotional bond. In PK experiments it's clear that if you don't use experts in this kind of task, you can obtain nothing; but if you enroll PK experts, you can obtain psi phenomena.

In a meta-analysis of forced choice experiments with people in normal states of consciousness when they're asked something like, "Try to guess which picture will be shown to you," there won't be an effect because it's necessary to have training and to select the participants. In remote viewing, you must decrease all your mental activity in trying to depict the scene, but untrained people may not be able to do. *Bill Bengston only picks skeptics to heal his mice and excludes anyone who is a psi believer.* Some psi phenomenon can be done by non-experts.

In terms of methodology, you push for open practices such as the Peer Reviewers' Openness Initiative where you put all your data online. Yes, it's part of the open-science revolution. Research materials should be available to other people in order to allow them to reproduce your findings. This is becoming mainstream and many scientific journals make it mandatory to make data available online. *What about the alternatives to the null hypothesis significance testing? (The rejection of a zero correlation or difference between two or more conditions.) You're asking for a different approach to statistical analysis?* Yes, this is one of the causes of the crisis of reproducibility. It's a statistical approach that is still used inappropriately in many scientific fields. We need new ways to analyze the data. Parapsychologists are not far from this kind of approach.

What are your favorite parapsychological journals? I read the *Journal of Parapsychology, Journal of Scientific Exploration,* and the *Journal of the Society for Psychical Research.* If you look at my publications, most of them are published in non-specialized parapsychology journals because it's a good way to reduce prejudice against psi phenomena.

As a productive person with over 90 articles, what do you do for fun and to nurture yourself? I investigate psi phenomena because it's very exciting. A lot of people are hungry to know more from a scientific point of view. Also I walk in nature, meditate, I lead a normal life. I married a teacher. My fatherly duties are now fewer so I have time for my interests.

The world situation is alarming with only a decade to take action about climate change. Do you consider yourself an optimist or pessimist about the world situation? I see many antidotes to the world crashing, because even in your country, there is an increasing number of people trying to pursue the spiritual way by doing meditation and mindfulness, the real antidotes. *Is there anything else that you would like to add?* I hope that our studies of the practical application of mental interaction from a distance and so on are replicated by investigators to convince people that our human potentialities are much more than we experience in normal life.

Book

La Parasicologia, 2007

Other publications: https://scholar.google.it/citations?user=uoURuxcAAAAJ&hl=it

Science of Consciousness Research Group: https://dpg.unipd.it/en/soc

Endnotes

1 Tressoldi, P., Pederzoli, L., Matteoli, M., Prati, E. and Kruth, J. (2016) "Can Our Minds Emit Light at 7300 km Distance? A Pre-Registered Confirmatory Experiment of Mental Entanglement with a Photomultiplier." *NeuroQuantology*, 14, 3.47-455. doi: 10.14704/nq.2016.14.3.906.

2 https://www.ncbi.nlm.nih.gov/pmc/articles/PMC2600606

3 https://www.facebook.com/groups/573425379393263

4 https://psyarxiv.com/s7uad/ https://github.com/tressoldi/

MindSwitch
5 https://youtu.be/-W6SZ1fKFeY
6 Bilucaglia, M., Pederzoli, L., Giroldini, W. et al. (2019) "EEG correlation at a distance: A re-analysis of two studies using a machine learning approach." *F1000 Research*, 8:43.
7 https://f1000research.com/articles/4-457
8 Tressoldi, P., Pederzoli, L., Caini, P., Ferrini, A., Melloni, S., Prati, E., Richeldi, D., Richeldi, F. & Trabucco, A. (2015) "Hypnotically Induced Out-of-Body Experience: How Many Bodies Are There? Unexpected Discoveries About the Subtle Body and Psychic Body." *SAGE Open*.
9 https://papers.ssrn.com/sol3/papers.cfm?abstract_id=3281560
10 Pederzoli, L., De Stefano, E. & Tressoldi, P.E. (2019) "Hypno-Death-Experiences: death experiences during hypnotic regressions." *Death Studies*.
11 Pederzoli, L., De Stefano, E. & Tressoldi, P.E. (2019) "Hypno-Death-Experiences: death experiences during hypnotic regressions." *Death Studies*.
12 Duma, G.M., Mento, G., Manari, T., Martinelli, M. & Tressoldi, P.E. (2017) "Driving with Intuition: A Preregistered Study About the EEG Anticipation of Simulated Random Car Accidents." PLoS ONE 12(1).
13 Cardeña, E. (2018) "The experimental evidence for parapsychological phenomena: A review." *American Psychologist, 73*(5), 663-677. https://psycnet.apa.org/doiLanding?doi=10.1037%2Famp0000236
14 https://www.youtube.com/watch?v=6rsmIcnmISc&t=126s

Russell Targ

Remote Viewing Teaches Us About Reality

Photo by Patricia Targ

Questions to Ponder

What traits and abilities characterize skilled remote viewers? How do they access information?

How does remote viewing indicate that our linear concepts of time are limited?

What experiences led Russell Targ to take psi phenomenon seriously? What convinced him that consciousness survives death?

Geometry explains psychic abilities, not quantum physics, according to Targ and Elizabeth Rauscher. Explain.

I was born in Chicago, Illinois, April 11th, 1934, an only child. *Do you think of yourself as an Aries?* Yes, Aries are enthusiastic about everything and they are especially enthusiastic about their birthday. The most enthusiastic experience I have right now is celebrating my birthday next week and finishing my film. I've been working five years on a documentary film about psychic spying and the CIA, called *Third Eye Spies. Are any of the original remote viewers* [*RVers*] *in the film?* Yes, Joe McMoneagle is probably the greatest RVer living today and he is in the film talking about his experiences working with the army and the CIA. I was an interviewer with Joe, in the first RV at Stanford Research Institute (SRI).

Have you found common characteristics among these talented viewers like Ingo Swann? They are energetic, intelligent, confident, and they have skills other than being psychic; none of these people were professional psychics. Hella Hammid was a much sought after distinguished photographer, Ingo was a wonderful visionary artist, Joe McMoneagle was a decorated senior warrant officer, and Pat Price was a retired police commissioner from the city of Burbank. *Then they don't have as much ego involved, it's not, "I've got to do this accurately because I am a psychic remote viewer."* That's exactly right. These people were confident and successful in what they were doing and intelligent. RVing is an intellectual activity, so you have to have your brain engaged to separate the mental noise of imagination and memory from the psychic signal. When you close your eyes and try and describe something, the image that comes into your mind does not have a tag on it that says, "This is brought to you by ESP." You have to learn to separate the diaphanous image from memory, imagination, guessing, naming and other distractive things-- that's a skill. They access information by quieting their minds and looking for the surprising images that appear in their awareness.

When we first went to work with the CIA, Sid Gottleib, the founder and director of the notorious CIA program called MK-Ultra in the 1950s, wanted us to give our viewers LSD because he thought that would make them more psychic, but I told him,

"That will certainly give them more pictures but it interferes with the single-pointed focus of attention they need." *And you were able to RV as well.* Yes, one time when Pat Price did not show up, I gave quite a good description; my excellent drawing shows that remote viewing is so easy that even a scientist can do it. You don't need special metaphysical preparation.

We'll circle back to RV but let's talk more about your childhood and your personality types. I'm an E/INTJ and a three on the Enneagram; I am a good engineer, able to focus on what I am doing and get it done. I spent 30 years as an engineer building lasers; that focus is a three on the Enneagram that the Buddhists call it single-pointed focus of attention. That's desirable if you are building hardware that's going to fly an airplane, my principal activity. I spent 10 years doing psychic stuff and the other 30 years we were building lasers.

I know from reading your autobiography that your father was interested in magic and science fiction and you did magic when you were an adolescent. I was an enthusiastic young magician indeed. I used to do magic on stage as a teenager. My father was a distinguished publisher in New York and a highly regarded intellectual in New York during my whole lifetime. He published *The Godfather* and the biography of Helena Blavatsky, the founder of Theosophy, so the idea of science fiction and metaphysics was certainly encouraged or permitted in the household.

As a teenager, I was doing magic on the stage where the magician pulls a piece of paper out of a hat and holds it to his forehead and says, "Somebody in the audience is looking for their lost cat," and a woman will hold up her hands and say, "Oh yes, can you tell me where Felix is hiding?" In the course of doing that kind of activity, I frequently got a picture of the person's home and their situation so I could supplement my stagecraft, by whatever ESP came my way. Since I was already cued up to the idea that there might be magic, I got very interested in ESP. Subsequently I talked to Milbourne Christopher, the famous

American magician, and the great mentalist called the Amazing Kreskin as well, and they agree that sometimes when a trick fails, they can supplement their magic trick with whatever ESP comes their way.

In high school, my classmate was Robert Rosenthal, who became a distinguished psychology professor at Harvard. One day in my sophomore biology class he showed up and lit my fuse. As a 14-year-old he brought us a deck of ESP cards. *The Zener cards used by J.B. Rhine?* Yes, our teacher gave us a day doing statistics applied to guessing the card image, and that launched me as a young teenager into reading the literature of ESP research. *What have been the most difficult times in your life and how did you cope?* The most difficult time in my life was definitely elementary school, because I went to school in the late 1930s and 1940s where even sitting in the front row I couldn't see the blackboard. I went through many years of school where the lessons were taught on the chalkboard and I couldn't see any of that so it was very challenging for me. But by the time I graduated from high school, I skipped two grades. I was 16 when I entered college, so the trauma of elementary school didn't really interfere with my education. But until very recently, I have had anxiety dreams about exams written on the blackboard.

What about Columbia? It must have been fairly difficult because you left after two years of physics grad school. Columbia was very difficult. I was not prepared for Columbia. I went to Queen's College in New York and had good grades and a nice interview at Columbia where they accepted me as a graduate student and a research assistant. They paid my tuition so I could do research with the famous Chinese physicist, C.S. Wu. But some of the theoretical classes were just way over my head. One of them was taught by T.D. Lee, the Nobel Prize-winning Chinese physicist, whose English was not very good and I couldn't see what he was writing on the blackboard. We did not have a textbook as he was making up our first year of theoretical physics out of his brilliant head, and I did not pass that. I should say the final examination from T.D. Lee had a passing grade of 30 out of 100 so I was not the only one who couldn't make any sense of the course. I did well in most of my classes in Columbia but there were some things that I couldn't do.

While other people were going out with girls, I had this little flock of four or five very smart, engaging Jewish guys from the Bronx who I hung out with and played cards and chess. Gerald Feinberg went to Columbia when I did, got his Ph.D. in three years and two years later became chairman of the physics department. I didn't realize how much smarter they were than I was when they led me to Columbia. *Not having a Ph.D. hasn't slowed down your career. You were able to do laser research and wind-shear for airplanes so in the long run it didn't matter.* Yes, I published more than 100 papers in laser research, plasmas and microwaves. And I built an amazing 1000 watt carbon-dioxide laser that was used to heat-treat locomotive cylinders.

Since I retired from Lockheed Missiles & Space Co. in 1995, I've had 25 years where my principal activity is reading books. I probably read twice as fast as my darling wife Patricia who has excellent vision. I was at the Unity Church in Palo Alto because Mark Allen, who was President of New World Library, was giving a talk on visionary publishing and I wanted him to publish a book that I had just written, called *Miracles of Mind*. Patricia came over and introduced herself. She is an artist and we became friends. She walked me up to see Mark Allen who thought my book was interesting and I was with a very pretty young woman, which always makes a nicer package. So at that instance in church, I found my wife to whom I have been married for 15 years and I sold my book.

Carl Jung says that relationships bring up unconscious issues so they can be challenged. I had been married to Joan Fischer Targ, my first wife, for 40 years until she passed away. Joan was the sister of the World Chess champion Bobby Fischer. Then I had a partner for a decade: Jane Katra was my writing partner and we traveled around the world teaching for a decade.

My mother and father were not religious Jews. In fact when I was 13 years old, I thought it would be good to be bar mitzvahed, but my father said, "No child of mine will ever be part of such a practice. Most of the suffering in the world is caused by organized religion." I was 13 right after the end of World War II so we were familiar with the German atrocities and many other religious atrocities, so my father strongly felt that he didn't want to be part of organized religion.

You were drawn to an eclectic religious background, the Course In Miracles *(which is supposed to be a new channeling by Jesus), Buddhism, and your teacher Gangaji from the Hindu tradition. How did that all synthesize for you?* I knew that there was a reward or opportunity for spiritual experiences. I got the idea when I was 20 and entered graduate school at Columbia. One of the women who worked with me was a member of the Theosophical Society in New York. I read the teachings and became interested in kundalini meditation which I did for quite a while, but I don't recommend that anyone do that without a teacher. You can have an energetic experience, but you can get into mental and physical problems. The Theosophical teachings prepared me for reading Buddhism which is more straightforward and less doctrinaire. Buddhism gives you powerful tools for dealing with the suffering around you, how to quiet your mind to parse the suffering and be able to distinguish pain from suffering. When people try and meditate and they can't get their mind quiet, it's because they have no experience finding the off switch. Buddhism gives you tools to do that.

What's an example of an off switch that worked for you in your meditation practice? I've been meditating for 40 years, so if someone asks me to find something for them that they've lost, I can swing my chair around, close my eyes and occasionally tell them where it is. I subscribe to the ideas of Spinoza and Einstein who felt that God is the organizing principle of the universe. When I am quieting my mind, or when I go to bed at night, I am happy to thank God or the universe for my good fortune that I was brought up in a highly-educated household and was given a pretty good brain. On the downside, I was given extremely bad vision and I have a non-coagulation bleeding problem so I almost died several times in the course of my life. In spite of that, I drove my motorcycle for 35 years and didn't get killed. I've gotten to be 85 years old and am in pretty good health. I am happy to give thanks for the organization of the universe that allowed me to be sitting here at my old age talking to you about the things that excite me.

If you were going to give advice to someone who said, "When I am 85, I want to be as bright and interested as Russell Targ," what would you tell that person? Unfortunately the answer is to choose good parents. *Does that mean good genes or good parenting?* That means good genes. *You also had colon cancer and maybe liver cancer. So you overcame that as well?* So it seems, yes. *Some people suggest that cancer and other diseases are metaphors, a signal from the body saying let's correct course here. Do you feel like that or do you feel it's purely a physiological function?* Your immune system deals with cancer. I was exposed to formaldehyde, gasoline, airplane glue, and other carcinogenic items. Some people will get cancer from them and some people won't, depending on your immune system. I don't know of anyone who says that meditation will keep you free of cancer. I think that if you keep your immune system strong, assuming you can figure out how to do that, then you are less prone to get cancer. Certainly if you cut down on your beef steak, that's probably good for your health.

Barbara Stone had breast cancer and she said in her therapeutic practice, she finds breast cancer often has to do with nurturing everyone else but not yourself. She went through a difficult divorce and it took 10 years for the cancer to surface. Did you have anything that was upsetting to your immune system before your colon cancer?* Yes, it was the end of the ESP program at SRI so I was in transition. I left the RV program after a decade after starting the program with my laser colleague Harold Puthoff in 1972. Hal and I were both known to NASA and the CIA, so we could go to these wealthy organizations and tell them, "We've done research with you in the past, it worked. We want some money now to investigate psychic ability." So we had a very well-funded program for a decade. At the end of a decade, it became more and more classified but I didn't grow up to be a psychic spy for the CIA. By 1982 we couldn't publish anything anymore of our Top Secret program because the CIA thought this is too valuable to release any further information. So I left the program. So you could say when I got cancer in the early 80s, I was in transition, but I was in an enthusiastic frame of mind. In 1983, we were forecasting silver commodities with RV and made $120,000 in the silver market.

Why didn't you keep on playing with the stock market since you had success with silver? Because it's a hit and miss proposition. The first time, we did nine trials in November/December of 1982 and were successful at all nine of them, which generated a lot of publicity. Our investor was so excited about making all this money that he wanted us to do viewings twice a week rather than once a week. But our protocols were if we did them twice a week, the viewer did not get feedback for trial number one until he or she had done trial number two, and that lack of feedback was a problem. We also got a little carried away in the original nine experiments as a scientific adventure, but as we began the second nine, it was how much money were we going to make. We went from a spiritual, scientific adventure into a money-making adventure, and that change of outlook might have affected our result. Other people have now used our protocol, called Associative Remote Viewing.

What do you think about the increasing number of NDEs recorded since we have better resuscitation techniques—what do they indicate about what happens after death? The NDE is controversial. People have experiences near-death but I am not sure that it pertains to dying, but I do think for the people who have truly died that their consciousness is still available--in that deceased people can actually be helpful and can give you information. During a tonsillectomy when I was five, I spent a long time under ether anesthesia. I was aware at the time that I was floating near the ceiling of the green operating room. I could see the stainless steel tray of operating instruments and heard all sorts of very weird noises. We have people in the laboratory who can describe distant things and have OBEs where they can see themselves lying on the ground without having an NDE.

I think there is very good data that deceased people can communicate. *After the tragic death of your brilliant daughter Elisabeth, she turned off lights to say hello. Did she communicate in any other ways with you?* Several of us were sitting on the deck of our house in Portola Valley looking over the San Francisco Bay, the day after Elisabeth died, talking about her. All the lights in the large house went out and then came back on. Her husband was there and we wondered, is that Elisabeth? Then they went off and came on again as though to answer the question.

Elisabeth has given messages to other people. I was the only living person who knew about an event she described. *She gave the messages in Russian to some people.* That's right. She had been doing healing experiments with AIDS patients. Yes, after she died, she appeared before a nurse in that experiment and said, "I want you to give a message to my husband," except she gave the message in Russian. Elisabeth was a fluent speaker and this nurse couldn't make any sense of it. Elisabeth parsed that into one syllable groups, saying, "Write down *ya,* write down *lyu,* write down *blyu.*" She gave two sentences, "See you" and "I love you." That was Elisabeth's fine hand setting up that brilliant experiment and proof from beyond.

What really impressed me is that Elisabeth gave somebody

a message saying, "Make sure that you give my father this message because that will convince him that I really survived because nobody knows this but him." That convinced me that she survived because it was an embarrassing event in early child-rearing only I and my wife and Elisabeth knew about and my wife is deceased. When Elisabeth was three years old, my mother sent her a dress to meet her at the airport when Grandma flew to California from New York. The problem was that Elisabeth never wore a dress and neither did my wife; they wore overalls or slacks. Elisabeth did not want to wear a dress so we sort of stuffed her into the dress like you stuff laundry into a laundry bag and she screamed her head off. She told us later that this was really traumatic for her but we were young parents who didn't know any better. She had been dead 10 years at that point, but she still wanted to give a message to me since I was much more skeptical than I am now.

Did you find that in your engineering work you had access to helpful information in addition to your logical left-brain? No, but about once a month I will have a precognitive dream that pertains to events that are going to happen the next day. What I've learned is that if I have a dream that is not an anxiety dream, or a wish fulfillment dream, or a dream based on the previous day's residue, but rather is a bizarre dream of unusual clarity, that's probably a precognitive dream. We live in a house with a high-pitched cathedral ceiling and in a dream I had an electric train running around the lower portion of the living room on the ceiling. I told my wife about that because it was a very clear and interesting dream that didn't pertain to anything. When I looked at *The New York Times* the next morning, on the front page was a reconstruction of the elevated rail in downtown Chicago where I grew up. It was exactly what I had described, an electric train running in a circle on an elevator track that was seen in my dream. My father's old bookstore was on that very street. I would say the picture in *The New York Times* was the cause even though it resided in the future.

I dreamed that you were driving a Zamboni cleaning off the ice field and then you were driving a lawn-mowing machine. I thought, "He is cleaning off these surfaces." I am very into polishing glass. Hanukah is coming up, so I was busy polishing my Menorah a week ago with silver polish and was really into polishing stuff—that's the sort of thing that a three on the Enneagram likes to do. Eli Jaxon-Bear, who is Gangaji's husband, is an enthusiastic teacher of the Enneagram so I like training with him. I spent a decade sitting with my principal living teacher, Gangaji. When people ask me who is your teacher, I usually tell them Longchen Rabjam, who is a 12th century Buddhist teacher who put together all the teachings of Padmasambhava, the eighth century Buddhist who came to Tibet to teach Buddhism. He wrote a book called *Self-Liberation Through Seeing with Naked Awareness* where he invited you to quiet your mind and experience who you are as timeless awareness. Then 400 years later Longchenpa wrote several books describing the process without the iconography of Hinduism, without the gods and goddesses.

How has Gangaji influenced you? She is a follower of Ramana Maharshi who died in 1950; his famous student H.W.L. Poonja taught Gangaji. The teaching is to spend some time to quiet your mind and see who is really there--*Advaita Vedanta* means no separation and Gangaji invites you to have the experience of no separation and the expanded awareness sitting with her. *You say in your autobiography that happiness is not craving, not wanting things, not needing. Does that follow from knowing who you are as a spiritual person?* Craving for what the Buddhists call cherished outcomes is a sure road to suffering.

To answer that question "Who am I?" what did you come up with? Using the language of Longchen Rabjam, my nature is timeless awareness. I am very experienced now in quieting my mind and moving into that spacious realm of timeless awareness. It's much like an invitation to RV, so when I ask a viewer to describe what their partner is hiding, I'll say, "I want you to close your eyes and quiet your mind and look for surprising images that appear in your awareness. Don't tell me where Joe is hiding, just tell me what you are experiencing. You can't be wrong. All I want you to tell me is the surprising images," and that turns out to be very easy to do. In my film *The Third Eye Spies* I show how we teach people how to do that exact thing so they find downed airplanes, hidden soldiers, kidnapped people, etc. The point of the film is that RV and psychic abilities are a natural ability. It's like showing somebody how to do an ability they already have rather than teaching, and if I set the stage then they can do it.

Critics say after the CIA funded the Star Gate program for over 20 years, that there were no results, that the RVers were given too many cues, Uri Geller used tricks, etc. yet they are so clear about specifics like accurately drawing a Russian missile silo. They lie. We open our film with Jimmy Carter saying, "The most amazing thing that happened to me in my presidency is we were looking for a downed Russian bomber in Africa and we couldn't find it because it had crashed in the jungle and the satellite photography couldn't see through the trees. We found a woman RVer in California who gave us the geographical co-ordinates. We sent a CIA group into the jungle in a helicopter and we found the plane just where she saw it." That's basically what we did for a living for 20 years. The CIA would come to us to find kidnapped people as when we gave a medical report on the kidnapped hostages from Iran, where one of them was very sick and we described his sickness. We said not to worry because we see him being taken out of the country on an airplane and that in fact happened two days later.

How can this be discounted when it's so specific? People are worried about being teased about being psychic, but in *Third Eye Spies*, we have our CIA contract monitors: Kit Green was a physician with the CIA and Ken Kress was a physicist undercover at the CIA. They were in charge of our program, senior scientists who looked into the camera and said, "Yes, we were polygraphed and what Russell is describing really happened. We were there. They RVed the transcript locked in the safe, they found the downed airplane." *There must be some reason that people are so vehement about denying evidence.* The Catholic Church doesn't like it; it's always worried about freelance spirituality. RV seems like a metaphysical activity. It's all right for you to talk to God but you get in trouble when God talks to you. At the time of the Enlightenment, we got mind-body separation, where Descartes says that the body is made of meat and potatoes and the mind, which will survive after death, is quite different. So the idea that your mind can travel around and give you information is contrary to the accepted ontology. Basically it's forbidden by the Catholic Church and also it's not readily explained by physics.

We get information from the future but causality says you can't see into the future. Now physics is becoming much more accepting of that. I presented at two recent conferences sponsored by the American Institute of Physics that dealt explicitly with retrocausality, meaning the physics that occurs in the present is caused by the future—it had become a much more accepted and important part of modern physics. *What's an example that a physicist would give of retrocausality?* There are a number of things that occur in particle decay where you have something caused by an event that appears to take place in the future.

Do you think that the Russians, Chinese and US intelligence agencies are still doing RV and using psychics, but disparage it because it's secret? When I was in Russia in 1983 to 84, there was great interest in psychic abilities at the USSR Academy of Sciences and in the time I spent in Yerevan, the Armenian Academy of Sciences was doing remote viewing that looked very much like ours. A senior psychology professor, Ruben Agazumsan, told me that he sometimes had his RVers describe targets where someone was going to be hiding and they described them before the target was even chosen. So they had their real time RV contaminated by precognition. Kit Green says on camera that the CIA is still doing this, to the best of his knowledge. If somebody says, "Why would you trust the guy in the laboratory who says some crazy thing?" The answer is you wouldn't. You would want two or three other sources as you always have in the intelligence community. *Right, replicability is what science is about.*

In your autobiography you quote physics professor Henry Stapp of UC Berkeley as saying "non-locality may be the most important discovery in all of science." Can it explain why we can do remote viewing and other paranormal phenomena? Yes, non-locality shows that we misapprehend the nature of the space and time in which we live. I agree with him in that non-locality is a very important idea. Erwin Schrödinger said that "consciousness is a singular of which there is no plural." He was an outspoken Vedantist; he felt that consciousness is primary. The accuracy and reliability RV are independent of space and time. It is a non-local ability, in line with the non-locality of modern ideas in quantum mechanics. The fact that you have non-locality makes it appear that there is a quantum mechanical effect where particles created at the same time all remain connected to one another like identical twins. I don't believe that's the explanation for psychic ability.

I think psychic abilities occur not through quantum mechanics but through the geometry of the space we live in. The mysterious phenomena of psychic abilities cannot be explained by any kind of electromagnetic field, because psi is independent of space and time, whereas fields have to obey the inverse square law of decreased energy with distance and they propagate only from the present to the future. In a mathematically complex geometry, there will always be a path of zero distance connection at any two points in the space-time manifold.

Famous physicist Archibald Wheeler said the answer "is not in the fields, it's in the geometry." We live in a multidimensional space-time where there is always a path or a trajectory through this space so there is no separation. *How many dimensions are there?* We live in a normal four-dimensional space-time with three space dimensions and one time dimension. My partner Elizabeth Rauscher published widely about the idea that each of those four dimensions has a real part and an imaginary part. What we normally see on the kitchen table is the real part, but there is also an imaginary part which is compatible with modern physics. Space-time is complex with the real and the imaginary. There will always be trajectory through complex space-time where the sum of the parts making up the distance adds to zero. That's how ESP works.

For example, if you were working in high school with the hypotenuse triangle it's x^2+y^2, the Pythagorean theorem. Now in complex space, if one of those is a real side and one is an imaginary side, then you could add the real x^2 to the imaginary $(iy)^2$ and if $x=y$, that would add to 0 because then x^2-x^2 is a hypotenuse in the eight space, they add the 0! Some people say, "Maybe ESP is in another dimension," but we are not adding another dimension just for the fun of it to explain ESP. We are saying that our present space-time could be described as a complex space-time.

In 1905 Hermann Minkowski developed the complex space-time that Einstein uses in Relativity Theory, in that he measured distances with three spatial dimensions and a complex time dimension. The square root of minus one times the speed of light times the time, ICT. If you square that, you get a distance squared. So in relativity, distances are $x^2+y^2+z^2+(iCT)^2$ where T is the time and that squared becomes a negative number. You have a complex space-time that works and relativity works all the time; it's been proven now for 100 years. So we're saying that Minkowski initially proposed that all those dimensions be complex while Einstein said, "I don't need them to be complex. It's enough for time to be complex." I emphasize that this is not some weird thing that we made up to solve the ESP problem but is a bona fide geometrical metric that does other things. Complex space-time, called complex Minkowski space-time, is a recognized concept.

Is probability another word for imaginary? Imaginary is a misnomer. You can't take the square root of minus one, so it was called an imaginary number but it's a perfectly good number. All electromagnetic theory and applications involve and require complex numbers. It's an unconventional way of writing the space-time that we live in that gives you zero total distance; it always gives you a path. There will be a path through space-time from where I am to where you are so that there is actually no separation between us. The geometrical description of psychic abilities is simpler, it's more parsimonious than trying to get a quantum mechanical explanation.

Does the fact that 95% of the universe is dark energy and dark matter relate to this imaginary complexity? No, that just shows how ignorant we are and that the universe is full of things we don't understand. Kurt Gödel was a great mathematician of the previous century who wrote very famous incompleteness theorems that show whenever you have a body of axioms describing something, it will contain some axioms whose truth or falsity you can't prove so there's fundamental uncertainty. Gödel is considered one of the greatest mathematicians of all time because he shows that there are limits to what we are able to prove.

Niels Bohr gave the first coherent description of the atom, what he called complementarity. Everybody wants to know is light a wave or is light a particle. Sometimes when it goes through a prism and breaks up into a rainbow, it looks exactly like a wave, but where it strikes the detector of a light meter it goes tick, tick, tick counting off the photons. It's a particle or a wave depending on how you measure, which is of course very unacceptable but it's the truth. So Bohr gave the answer which has now been accepted for 75 years that it has a dual nature; light is both a wave and a particle. Or better, light is both a wave and not a wave.

The Buddhists understood this. At the time of Christ there was a great grammarian and philosopher named Nagarjuna who said that most things are neither true, nor not true. A lot of the suffering that we experience is because of Aristotle who 500 years before Nagarjuna said that the middle is excluded; a thing is either white or it's not white. There is no gray for Aristotle. He said that duality makes for a lot of suffering, issues like the so-called mind-body problem, for example. Is this a mind or is it a body or is it separate? Nagarjuna would say they are neither one nor separate, it's the wrong question. The mind-body problem has no answer because it's a false question. Nagarjuna would say, just like the body, light is neither separate nor united, there's only one thing with two manifestations. I think psychic functioning is in that realm.

As we learn more about the eight space complex geometry we will see more into the future. It involves eight-dimensional complex space-time, first proposed by Minkowski who proposed it to Einstein as related to the complex of space-time of General Relativity. You have two points in space-time. One is where we are and say the other one is the San Francisco Warriors basketball game in Texas tomorrow. So there are 1,000 miles and 24 hours between me and that basketball game but in this model of space-time, there will be a path through space-time that goes from here and now to tomorrow and Houston so we can see the outcome of that basketball game, just like forecasting silver futures. I could close my eyes and meditate and see the final score of the game tomorrow because that path is available.

Does the path only exist if you observe it? No, all paths are available because we live in a connected world. The great secret in the ESP and the RV world is that it's no harder to describe something in Soviet Siberia than it is to describe something across the street. Increasing the distance does not at all make it difficult and looking into the future is no more difficult than describing something current. Our standard experiment is somebody goes to hide in a distant location and then I guide the viewer to describe where Joe is hiding. In addition, I can ask you, here we are at 12:30, tell me where Joe is going to be hiding at 2:30 even though he has not found his place yet. That works just as well or perhaps better; the evidence is that precognitive RV is absolutely as successful as real-time RV. So the path we have through space-time through the hidden person right now or the hidden person two hours in the future, is as easy to do. The accuracy and reliability of RV are independent of space and time. It is a non-local ability, in line with the nonlocality of modern ideas in quantum mechanics. We've published those findings in the premier journals and that work has been replicated.

It makes us realize that we understand a tiny percent of knowing reality. The fact that we can see into the future as well as into the present supports the idea that the complex space-time model is the explanation rather than quantum mechanics. Quantum mechanics allows teeny-weeny things to show reverse causality effects microscopically. RV done by competent people and experienced viewers works most of the time; when we sat down to forecast silver we did that nine times in a row describing what's going to happen in the commodity exchange 50 miles from where we were and five days in the future. It was on the front page of *The Wall Street Journal.* RV worked very well in our lab for 20 years, so it's a mistake to think RV is a weak effect. In the hands of people who actually know what they are doing, it's very reliable because we were in the army program. I trained six people for the army Psychic Corps who went on for a decade doing application for the CIA and they scored at odds of one in a million in our lab during training.

Let's look at the books that you've written. Limitless Mind *is mostly about RV?* Yes, that was my first book about RV and it includes some of the work that my daughter Elisabeth did in distant healing. My last book was called *The Reality of ESP: A Physicist's Proof of Psychic Abilities.* It describes the history of our work with RV and we have a chapter telling people how to do it working with a friend, how to develop your own psychic abilities. So, for a person starting to understand and use psychic abilities, that book would be a good place to start.

What's next? I may make a fictional film about an RVer. *What about books you'd like to read in the future?* I am going back to read the Buddhist teaching of Longchenpa, the great teacher who told us that our nature is timeless awareness. He is a brilliant philosopher and a potent remote viewer. His famous twelfth century book *The Basic Space of Phenomena* reads like a physics text on the nature of consciousness, free of the laws of cause and effect; it is outside of time. For example, he says that because of our inherent timeless awareness, our consciousness isn't limited by cause and effect. So cause and effect is not an attribute of mind because mind is outside of time. Physicists were always shocked to know that you don't have causality. Physicists say if I don't understand causality, I don't understand anything and that's really a threat, and Longchenpa understood that. He very specifically says there is no cause and effect for consciousness because consciousness is outside of space and time, which is quite a daring thing to say in the 1200s. It sounds like contemporary relativity but he is talking about space and time, which he got by direct transmission.

Books

Limitless Mind: A Guide to Remote Viewing and Transformation of Consciousness, 2004

Do You See What I See? Memoirs of a Blind Biker, 2010

The Reality of ESP: A Physicist's Proof of Psychic Abilities, 2012

Targ, Russell & Puthoff, H. *Mind-Reach: Scientists Look at Psychic Abilities*, 1977

Targ, Russell & Harary, K. *The Mind Race: Understanding and Using Psychic Abilities*, 1984

Targ, Russell & Katra, J. *Miracles of Mind: Exploring Nonlocal*

Consciousness and Spiritual Healing, 1988
Targ, Russell & Katra, J. *The Heart of the Mind: How to Experience God Without Belief*, 1999
Targ, Russell & Hurtak, J.J. *End of Suffering: Fearless Living in Troubled Times... or, How to Get Out of Hell Free*, 2006

Jessica Utts, Ph.D.

Statistical Evidence for Remote Viewing

Photo by Brian Phillips

Questions to Ponder

What do statistics show about psi phenomena such as remote viewing and ESP Ganzfeld studies? What other evidence got Dr. Utts interested in remote viewing? What kind of evidence is most compelling to other statisticians she addresses?

What's an explanation for why remote viewers were as accurate in describing a target that hadn't yet been selected? What questions does this raise about our notion of time?

I was born in Niagara Falls, New York in 1951, a Libra and ENTP. *What was there about your family that encouraged you to be excellent in math?* My parents were not mathematicians at all; my mother was a social worker, my father was a journalist, and that may be where I got my passion for promoting statistics to the public. But my mother was very supportive of me pursuing higher degrees and wanted me to do well academically, and that had a big influence. I think I led the life she would have liked to lead if she had lived in a different time period. She tried three times to get a master's degree and finally got one the same time I got mine. The first two attempts were thwarted by marriage and then a child. She was a strong encouragement to me and a great role model. I am second oldest of five children with an older sister, younger brother and two more girls after that. *Did they all gravitate towards high achieving careers like yours?* Not all. One of them unfortunately committed suicide, so she never quite got there. She was the youngest. My brother has his own business and he gravitated toward a science career and created a business out of that. My other two sisters became social workers like my mother, and one of them then created a business helping people declutter. Her social work training is helpful in that business, especially for clients who are hoarders.

There were subtle influences in my life that led me to be interested in math and so as an undergraduate major in college, I decided to be a math major. When I got to college, I started as a math major, and although I was doing fine it wasn't that much fun compared to what I saw my friends doing who were mostly psychology majors. That sounded a lot more fun to me so I decided to do a double major in math and psychology. The only course I had that overlapped the two majors was statistics. I had no idea what statistics was at the time but it seemed like the one thing I could do that would combine my two interests, so I decided to get a Ph.D. in statistics. At the time I didn't realize that the psychology part of statistics was just a small corner that wasn't what I would be expected to do in graduate school.

I was one of the few women in math and statistics when I went to Penn State for my graduate degrees. When I started my master's degree, there were several women in the graduate program with me, but at one point, I looked around in one of my classes and noticed I was the only woman left. I was the only one who went on to get a Ph.D. In fact I didn't realize it at the time, but I was the first woman to get a Ph.D. in the Statistics Department at Penn State. I don't think I understood then how unusual it was for a woman to be getting a Ph.D. in a technology-related field.

It doesn't sound like you felt isolated or that you were treated differently. It sounds like you were kind of oblivious. Oblivious is a good word for it. If I was treated differently it was because I was unique, so I wasn't just lost in the crowd. My professors knew who I was. I got along with everybody and it seemed to me, if anything, it helped me at that stage in my career to be the only woman in the Ph.D. program. I am sure there were plenty of times in my career when it didn't help that I was a woman, but at that point during graduate school, I felt like I was accepted and it was a good fit. I suppose if I hadn't done well, the professors might not have treated me as well.

Do you see a change now in undergrad math and statistic students in terms of gender balance? Yes, I definitely do. I think especially in statistics, unlike some other fields, undergraduates are more than 50% women nationwide. Statistics tends to be an area that attracts more women because it involves skills that can be used to make a difference and women seem to be more drawn to that, and to the idea of working in teams with other people collaboratively. I think critical mass makes a big difference too. As oblivious as I was, most women are not oblivious to who else is around them. And so I think the fact that there are plenty of other women in statistics and more women faculty members as role models attracts women to the field.

Aren't there organizations that are geared towards getting girls interested in STEM topics even in elementary school? Yes, in statistics, I think one important development that started in the late 1990s is the Advanced Placement Statistics (APS) program, in which high school students take a year-long course in statistics with a standardized exam at the end. If they do well on the exam, they get college credit for the course. We've seen an incredible increase in the number of statistics majors at the undergraduate level in the last 10 years or so, and I think quite a bit of that increase is due to the AP Statistics course. Students come into college aware of statistics as a potential major and profession. I was involved in the AP Statistics program for a very long time, including leading the grading for the last five years. It's very rewarding to see the growth in popularity of statistics in high school and college.

How did you get involved in doing the analysis of remote viewer (RV) and paranormal ESP kind of experiments? When I got tenured at the University of California, Davis, which is where I was for much of my career, I decided it was time to start branching out and look for things that I could do that were frankly more interesting than the theoretical statistics work I had been doing. I took a sabbatical leave at Stanford University and coincidentally I met Hal Puthoff and Ed May who at that time were running the RV program for the government at SRI (Stanford Research Institute), near Stanford University. They needed statistical help so I started working for them as a consultant and got really drawn into the work, partly because there were no other statisticians doing it. It's easy to be the world expert when you are the only one doing something! I started attending parapsychology conferences and met other people in parapsychology and one thing led to another. A few years later, I took a year of leave of absence from Davis and worked at SRI as a visiting scientist and that kind of clinched my interest, because I got to see hands on day-to-day what was going on.

You report in some of your articles that there's not a huge difference in the ESP experiments, like chance will be a 25% hit rate but then the average is 27%. So I was surprised that the rate was so low, including for RV because they seem to be amazingly accurate. The 27% refers to forced choice experiments (such as guessing the suit of a card) where usually there is a one in four chance of getting it right by chance. Free response experiments such as

RV often are designed to have a 25% hit rate by chance as well, but the experiments are much more complex to do than forced choice experiments. But with RV and a similar type of free-response experiment called Ganzfeld, we see more like a 33% hit rate. That's pretty convincing and especially when you see it over and over again across laboratories and experiments, in thousands of trials across modalities including Ganzfeld and RV. Ganzfeld experiments are a little bit more complicated than RV but basically they measure the same thing. They include a "sender" who is looking at the target while the "recipient" tries to describe it. At the end of the viewing in these experiments, the recipient (the person trying to be psychic) is shown four choices and asked which one they thought the sender was looking at. As with RV, the correct answer is randomly chosen from a set of four possibilities but those four are chosen from a larger set. Consistently choosing the correct target about 33% of the time instead of the 25% that would happen by chance provides convincing statistical evidence.

But those statistics don't capture what I think you are talking about, which is that occasionally we see some striking results when the participant almost perfectly describes a hidden photograph or location that they have no way of "seeing" through normal means. The experiments are set up to have very conservative statistics. In advance, four possible target choices are identified, and one is picked at random (using an RNG) to be the correct answer. If the response provided by the RVer is good enough so that somebody could pick out the right target from the four possible choices, that scores as a hit. It only has to be good enough to distinguish the right target out of the four choices, which are chosen to be as dissimilar as possible. If the RVer provides a perfect description of the target making it an exact match, it gets no more credit than one that's just good enough to pick the right one out of the four, so the statistics are conservative in that respect. We've tried various other methods like trying to list all the possible features that could be in a target and in a response and match them up but nothing seems to work quite as well as a human judge (blind to the right answer) looking at the RV response and seeing which picture of the four possibilities matches it the best.

I am sure you tried RV. Did you succeed? I've only tried a few times and never in a formal experiment. If you ask Russell Targ,* he'll say I succeeded because he led me through an informal RV during dinner at a restaurant. He thinks I did a really great job. I feel a little bit more critical and not as convinced and it wasn't a formal experiment. And it was just a single session. *The procedure is to get into the RV frame of mind, you quiet your mind. How do you do that?* It depends on the experiment. At SRI we were mainly using the same people over and over again because it was a classified program, so they each had their own methods of relaxing. In general, what's done is a relaxation exercise before the RV or Ganzfeld experiment. *Could that be something as simple as your feet are relaxed and heavy, your ankles are warm and so on?* Right, the body-scan relaxation exercise. I am going to be doing a demonstration pretty soon for a group that asked me to come and talk about RV, which will be my first time to lead people through a RV on my own. I plan to use a four-minute relaxation exercise *The New York Times* provided in an article about meditation.[1]

The statistics are what really convinced me that RV is possible, but what intrigued me early on was the results when my father participated in a Ganzfeld experiment at a lab in Princeton. At that time the lab was doing an experiment where they had people come in four times to see if the same people could do a repeat good or bad performance. In this experiment there were a few hundred possibilities from which the four targets were chosen. I was the sender twice when my father was the person being tested. All four times my father went he got a first-place match but the two times that I was the sender, he not only got a first place, he did one of those things where there's no question that he got it right. The first time my father participated, when I was not the sender, the correct target material was a picture of the Kremlin. The second time he went, I went with him to be the sender, and on the way there he described how the target the first time was the Kremlin. I was quite surprised when I went into the sender's room, and out of all of the possibilities, the Kremlin had again been randomly selected as the target.

During the session the sender can hear the receiver trying to describe the unknown target. My father said, "Oh I see red and I see gold," and then he said, "It's the Kremlin again. I know it shouldn't be, but it is." And he was right on. So of course when my father was shown the four possible choices at the end, and he saw the picture of the Kremlin, he said, "That's it." *And your job was to just focus your intention on the image?* Exactly. And try to send it to him in some way. The second time I went as a sender for him, the target turned out to be a short video clip of a spaceman in a white spacesuit floating against a blue background. My father started talking about the painting called *The Blue Boy,* the famous painting of a boy in a blue suit with a white background. So when he saw the four choices, as soon as he saw the one of the guy floating in space, he said, "That's it. The blue boy." So again, a pretty good first place hit. He went one other time with a different sender and again he got a first-place hit. Around that time I was thinking, "All this could be fraud and cheating," I had only seen RV at SRI. They used the same people over again and I thought, "Maybe they are in cahoots; they are just trying to convince the government there's something there." But I knew my father wasn't cheating, so that was something that helped to draw me in even though that's not nearly as convincing as hundreds and hundreds of studies that show strong statistical results.

While you were at Princeton, were they researching with the RNGs? This was a different lab led by Charles Honorton who is the one that really started using the Ganzfeld technique. His lab in Princeton was called Psychophysical Research Labs but it wasn't on the Princeton campus.

After 20 years, the government's conclusion was that Star Gate hadn't produced anything useful, there were too many cues given to the viewers and people like Uri Geller did sleight-of-hand. Why do you think there was such dismissal of the results even though they are so concrete with drawings of the Russian silo or the new submarine, the lost airplane, and more? I think it's really important to separate out the two parts of the program. The scientific program was different from the operational RV program where they were trying to spy on our enemies. The early part of that operational program produced some of the evidential stuff that you are talking about. The difference is that in the operational viewing paradigm we can't assess how good the results would be just by chance alone.

For instance, let's suppose an important US diplomat has been taken hostage in a particular country. If a remote viewer is working on operational problems for our government, they might guess that they would be tasked with trying to describe where the hostage is being held, even though they were never told what problem they were trying to address––it was all coded in a blind way. But if they were following the news, they probably would figure out that they would be asked about the hostage, and might be able to guess the conditions under which he or she was being held and come up with a pretty good description just by chance. So it could look like a good match when it was really just guessing based on knowledge of the news. That's why the scientific program is more convincing to me.

I reviewed the scientific part of the program for the government in 1995, when they declassified the whole thing, along with skeptic Ray Hyman. We concluded that there was something going on statistically that could not be explained by chance. Ray, being a skeptic, said, "But it can't possibly be paranormal, there has to be a normal explanation." When I said, "Okay, give me one," he couldn't. Our report basically concluded that there was something established scientifically and statistically that was not explainable. However, we were not allowed to look at any of the operational materials and that was a bone of contention at the time.

Ed May, who directed the whole program, was really upset with good reason because apparently nobody looked at the operational stuff. *The operational is spying on Russian military establishment and the other part is doing experiments in the lab. What do those experiments focus on? I go to this location and then you draw where I am?* Right, except after a while they stopped doing "outbound" experiments and started just using targets that were either photographs or short clips from movies because it didn't matter if someone was actually out there. We would choose four potential targets that were as dissimilar as possible. (Sometimes five potential targets were used, but to keep it simple I'll describe what happened when there were four.) The person would do a RV by drawing and writing a description of what he or she thought the target was, and then a judge would look at the four possibilities, one of which had been randomly chosen to be the right answer. Choosing the target randomly is necessary so that there is no human bias built into the choice of the answer. And it's important to point out that the remote viewer didn't know what the four choices were.

The results showed that when there were four choices, about one in three times, the judge could identify which one had been randomly chosen to be the target, just by looking at the RVer's description. So again that's not great but statistically it's not bad either, especially since some of them were really close matches and descriptions. There were other experiments we tried to do like actually locate a missing person or material. In one operational case, the RVers were trying to locate a military aircraft that crashed in Africa before the Soviets did. President

Jimmy Carter publicly talked about how we located it using RVers. But those are anecdotal so, as I explained earlier using the hostage example, there is no way to evaluate those results, compared to what should happen by chance.

My guess is the CIA wanted to shut the RV program down. It had been housed with the Defense Intelligence Agency (DIA) and they wanted to get rid of it. They supposedly wanted to give it to the CIA, but the CIA had enough of their own problems at the time and the Cold War was over. The CIA didn't really want it, so I think some influential members in Congress who wanted the program to continue were the ones that pushed to have this report commissioned. But I think those who didn't want the program to continue deliberately set things up so that they could make whatever conclusion they wanted, by putting the skeptic and me both on the committee. They knew previously what both of us had said about the possibility that RV works. By simply ignoring the operational stuff to a large extent, they concluded that it wasn't worth continuing, or so they said. Who knows? Maybe it is continuing.

To reiterate, the reason the operational material doesn't constitute proof is that to have a good experiment you have to prevent cheating or other means of ordinary communication and you need to know what should happen just by chance. Otherwise you can't compare what did happen to what you would expect to happen just by chance, and it's possible that someone did well because they had prior information about the situation. *In Russell Targ's book, he says that they located Patty Hearst when she was kidnapped.* Right, yes. That's an intriguing story. It's too bad the police didn't choose to follow up.

Could quantum physics, non-locally, explain the difference between a 33% actual and a 25% probable odds of getting those right answers? I am certainly no physicist but my thinking about it as a layperson is that there is probably no predetermined future. There are probably potential futures, some of which are more likely than others. In quantum mechanics, we know that there are various states that could be actualized with different probabilities. Consciousness comes in and actualizes one of them. Precognition, for example, could never be 100% accurate because I think what people are probably tapping into is the most probable future, not the one that's definitely going to happen. And so I think that could be partly what's going on here. The RV experiments seem to work equally well if they are done as precognition, when the target has not even been chosen when the RVer is trying to describe it. That creates a cohesive explanation for what could be going on because it could be that it's all precognition. Eventually viewers are going to be shown the right answer so maybe they are peering into their own future to see the target that they will eventually be shown. If we figure there's no predetermined future but just probable ones, then maybe some people are good at tuning into those probable futures.

In his interview Russell Targ talked about the eight dimension geometry that allows for a pathway into the present that allows for information flow. String theory and other theories talk about how we might live in a ten-dimensional world or maybe an eight-dimensional world but we only see three dimensions plus time. I can imagine that if that's true and we are projected into a three-dimensional world, that if somebody was somehow able to expand into that larger dimensional world, they might be able to see things that the rest of us can't normally see. As an analogy, there's a wonderful although sexist book written in the 1920s called *Flatland*. I think that's a really good analogy. It talks about how the people in *Flatland* live in a two-dimensional world and the hero is able to escape into three dimensions. He sees what we would see as the three-dimensional world and he comes back and tries to explain it to people, but they lock him up as crazy. It could be that something like that is going on here; that time and space are not as cut and dried as we think they are, three dimensions plus time, moving in one direction.

Some theories postulate that time is really simultaneous and that if you could sort of pop up above the three dimensions plus time that we live in, you could actually see what's going to happen. *Maybe that's what mystics do.* That to me is one of the most fascinating things about this area. I am totally convinced statistically that there is something going on here, but we just don't know what it is. I don't think we are going to convince anybody who is not already convinced until we come up with an explanation from physics or from a multidisciplinary team working on the problem.

Quantum non-locality explains it to me but Russell Targ pointed out that that doesn't work for precognition into the future. Right, non-locality would have to include time; you could have non-local time as well as non-local space. Suppose all time does exist at once, just like all space exists at once. So there's stuff going on right now in Australia but I can't see it because I am not there, so maybe there's also stuff going on right now in 2100, but I can't see it because I am not there. So it's possible that non-locality would include time as well as space. *In Robert Monroe's biographical books he astral travels to the future. He also goes into another dimension where it's part of him but in another body and he enters into that body and does things that are inappropriate. He realizes he should never do that again. In that dimension are different physics, the locomotion system was different, so his experiences read like science fiction, but his experiences indicate that the idea that time is accessible from various aspects might be true.*

How has this all impacted your own spiritual beliefs and your understanding of the meaning of life? I was lucky to not be raised in a religion that gave you "shoulds," I was raised in a Unitarian church. I feel very fortunate about that because the whole upbringing was that we should consider all different religious perspectives and ideas and make our own minds up about what we think is true. For me, that process is an ongoing life experience and I still don't know what I really think is true. I lean toward the idea that there is an interconnectedness among everything, not any kind of supreme being, but an interconnectedness and we all play a part in that. If that's correct, then you actually can change the world from your own living room by tapping into that interconnectedness and promoting the kinds of things that you would like to see in the world.

Is consciousness another name for interconnectedness? I think we each have our own individual consciousness as well. You can think of it like an octopus with many legs. So, all is interconnected at some level but then also we each have our own consciousness and individuality. *Hindus have said that for centuries in the form of individual Atman and overall Brahmin.*

What are you researching about now? I know you have textbooks about statistics. Right, I've written three statistics textbooks and my main goal there is to get people to understand statistics in a way that is useful for their daily life. The parapsychology research is really interesting because I think it's a good example of belief versus data. I've given a lot of talks to university audiences where I challenge people on this because the data is so strong that there would be no question that there is really an effect there, if it were any other non-controversial area. People's beliefs are very strong, and so their beliefs overwhelm the data in a lot of cases. There is a formal area of statistics called Bayesian statistics that incorporates belief and so I've been working on showing how to analyze parapsychology data using Bayesian methods where you can see why people can agree to disagree. If you have very strong beliefs then basically no amount of data is going to convince you, so that's been a really interesting way to approach the topic.

Can you statistically deal with the concept of people's beliefs? Yes, you can because you can quantify beliefs by asking the right questions. Let's use the word psi as a general term for paranormal abilities. The first question would be, "What's the probability that psi is possible at all?" People might say something like one in 50 billion, so you can quantify that. The second question is, "Let's suspend that belief for a minute and suppose it is possible, what magnitude of effect do you think would be possible?" So, for instance, if chance is one out of four, "What do you think it would be instead of that?" You can quantify that as well and put all that together with the data and come up with what's called the posterior probability. If people start with a small enough probability to begin with, then the posterior probability is still going to be pretty small. But if they start with a maybe 50-50 chance, then the data should convince them.

What do you find in a kind of meta-analysis of the groups that you've talked with? Do people tend to be skeptics or open? I've noticed a change over the years as people have gotten more open or maybe they are just more willing to admit it. But I think they are not convinced by data and it's kind of ironic to me that statisticians in particular, who are mainly the audience I talk to, claim that data is the most important thing. And yet, I'll ask them which would be more convincing to you, another 50,000 trials showing these same kinds of results, or one overwhelming personal experience? They all say one overwhelming personal experience, which I understand, but it's still kind of funny to me that even professional statisticians don't pick the data choice. I think that's part of human nature that people ignore data that doesn't support their beliefs and they exaggerate data that supports their beliefs.

What other psi studies do you point people to where there is above probability results besides RV and the other work at SRI? In the Ganzfeld studies, they are shown four choices and they have to pick which one they think is the target. It's a lot like RV and when I did my report for the government in 1995, I combined the data from those studies with the data from RV studies because I thought they were similar. Those studies were done in a lab at Princeton (the one where my father participated in experiments) and they were also being done in a number of other places, such as at Edinburgh University in the UK, at the Rhine Institute in North Carolina, and the University of Amsterdam. I combined all those results in the meta-analysis when I did my report.

What other universities study the paranormal? Arthur Koestler endowed a chair in parapsychology at the University of Edinburgh that is still going on. That program produced a number of Ph.D.s who started programs at other universities throughout the UK, so that's where there is the most research going on. There isn't much in the US because unfortunately the US skeptics or deniers have been so successful in making it non-legitimate as an area to study. I was really lucky as a statistician, because we study data, so I wasn't actually doing parapsychology experiments, I was analyzing them. I escaped a lot of the criticism that I think I would have had if I had been doing the experiments.

Where is your future research taking you? I've become interested in ethical issues in data science, with the rapid increase in machine learning and artificial intelligence. With easy access to tremendous amounts of data I'm afraid companies will misuse it and make decisions without considering the quality of the data. For instance, machine learning algorithms that are used to make hiring decisions sometimes use data from past hiring and performance. But those previous decisions were made by humans, and often have gender, age, ethnicity, and other biases built in. Those biases then get perpetuated in the algorithms, but because the decisions are made by a computer algorithm, people think they must be valid. Statisticians understand how to assess the quality of data, and also how to report results in an unbiased way, so I've been giving talks and writing papers encouraging statisticians to get involved in data science ethics. I've always been active in statistics education and especially statistical literacy and trying to educate the public so I think I'll do more of that work as well. In fact that ties into the data science ethics theme, because people are less likely to be fooled by poorly developed and poorly reported statistical results if they understand some basic principles about data quality and statistical reporting. So I've been encouraging statistics educators to think about ethics and to focus on helping students understand the pitfalls.

Anything else that you would like readers and listeners to know about the statistics of the paranormal studies? I want people to realize that there are legitimate ways to study psi statistically and that when those methods are used the results provide convincing evidence. I think people also need to be a little critical of anecdotes. I have my own anecdotes that I could explain as chance or I could think they were paranormal, and I've seen extremes of belief on both sides. I've seen complete deniers and I've seen people who are too easily fooled by things that could easily be explained in other ways. So what I want people to know is there is an in-between ground, which is you can study this field statistically, you can design experiments to study it. My great hope and wish is that the stigma of doing so goes away so that we can figure out what's really going on here--because there is something going on. We are only going to figure it out if people are free to study it.

Books

Statistical Ideas and Methods, with Robert Heckard, 2005

Mind on Statistics, 5th edition, with Robert Heckard, 2014

Seeing Through Statistics, 4th edition, 2014

Endnote

1 https://www.nytimes.com/guides/well/how-to-meditate

Conclusion

We know very little about reality since visible matter is only about 5% of our universe and there may be many universes. This suggests the need for humility and open-mindedness about quantum physics and psi research that indicates on a subtle dimension there is no time or space, as well as universal connection or non-locality. As we move beyond the limited view that the only reality is physical matter that can be known with our five senses, we can expand our understanding of our abilities to perceive, to heal, and to care for our planet. We need university courses along with funding to do more research into the mysteries of what Einstein called subtle energy and Max Planck called consciousness. Our children need to be taught how to access their potential as humans with more understanding of reality. As Einstein said, "The important thing is to not stop questioning. Curiosity has its own reason for existing." Please comment and stay current on the book webpage.

Brief Peeks Beyond

Critical essays on metaphysics, neuroscience, free will, skepticism and culture

Bernardo Kastrup

An incisive, original, compelling alternative to current mainstream cultural views and assumptions.

Paperback: 978-1-78535-018-4 ebook: 978-1-78535-019-1

Framespotting

Changing how you look at things changes how you see them

Laurence & Alison Matthews

A punchy, upbeat guide to framespotting. Spot deceptions and hidden assumptions; swap growth for growing up. See and be free.

Paperback: 978-1-78279-689-3 ebook: 978-1-78279-822-4

Is There an Afterlife?

David Fontana

Is there an Afterlife? If so what is it like? How do Western ideas of the afterlife compare with Eastern? David Fontana presents the historical and contemporary evidence for survival of physical death.

Paperback: 978-1-90381-690-5

Nothing Matters

a book about nothing

Ronald Green

Thinking about Nothing opens the world to everything by illuminating new angles to old problems and stimulating new ways of thinking.

Paperback: 978-1-84694-707-0 ebook: 978-1-78099-016-3

Panpsychism

The Philosophy of the Sensuous Cosmos

Peter Ells

Are free will and mind chimeras? This book, anti-materialistic but respecting science, answers: No! Mind is foundational to all existence.

Paperback: 978-1-84694-505-2 ebook: 978-1-78099-018-7

Punk Science

Inside the Mind of God

Manjir Samanta-Laughton

Many have experienced unexplainable phenomena; God, psychic abilities, extraordinary healing and angelic encounters. Can cutting-edge science actually explain phenomena previously thought of as 'paranormal'?

Paperback: 978-1-90504-793-2

The Vagabond Spirit of Poetry

Edward Clarke

Spend time with the wisest poets of the modern age and of the past, and let Edward Clarke remind you of the importance of poetry in our industrialized world.

Paperback: 978-1-78279-370-0 ebook: 978-1-78279-369-4

Readers of ebooks can buy or view any of these bestsellers by clicking on the live link in the title. Most titles are published in paperback and as an ebook. Paperbacks are available in traditional bookshops. Both print and ebook formats are available online.
Find more titles and sign up to our readers' newsletter at
http://www.johnhuntpublishing.com/non-fiction
Follow us on Facebook at
https://www.facebook.com/JHPNonFiction
and Twitter at https://twitter.com/JHPNonFiction